Toxicity Testing for Assessment of Environmental Agents
Interim Report

Committee on Toxicity Testing and
Assessment of Environmental Agents

Board on Environmental Studies and Toxicology

Institute for Laboratory Animal Research

Division on Earth and Life Studies

NATIONAL RESEARCH COUNCIL
OF THE NATIONAL ACADEMIES

THE NATIONAL
Washin
www.nap.edu

D1307203

THE NATIONAL ACADEMIES PRESS   500 Fifth Street, NW   Washington, DC 20001

NOTICE: The project that is the subject of this report was approved by the Governing Board of the National Research Council, whose members are drawn from the councils of the National Academy of Sciences, the National Academy of Engineering, and the Institute of Medicine. The members of the committee responsible for the report were chosen for their special competences and with regard for appropriate balance.

This project was supported by Contract No. 68-C-03-081 between the National Academy of Sciences and the U.S. Environmental Protection Agency. Any opinions, findings, conclusions, or recommendations expressed in this publication are those of the authors and do not necessarily reflect the views of the organizations or agencies that provided support for this project.

International Standard Book Number 0-309-10092-5 (Book)
International Standard Book Number 0-309-65652-4 (PDF)
Library of Congress Control Number 2006923288

Additional copies of this report are available from

The National Academies Press
500 Fifth Street, NW
Box 285
Washington, DC 20055

800-624-6242
202-334-3313 (in the Washington metropolitan area)
http://www.nap.edu

THE NATIONAL ACADEMIES
Advisers to the Nation on Science, Engineering, and Medicine

The **National Academy of Sciences** is a private, nonprofit, self-perpetuating society of distinguished scholars engaged in scientific and engineering research, dedicated to the furtherance of science and technology and to their use for the general welfare. Upon the authority of the charter granted to it by the Congress in 1863, the Academy has a mandate that requires it to advise the federal government on scientific and technical matters. Dr. Ralph J. Cicerone is president of the National Academy of Sciences.

The **National Academy of Engineering** was established in 1964, under the charter of the National Academy of Sciences, as a parallel organization of outstanding engineers. It is autonomous in its administration and in the selection of its members, sharing with the National Academy of Sciences the responsibility for advising the federal government. The National Academy of Engineering also sponsors engineering programs aimed at meeting national needs, encourages education and research, and recognizes the superior achievements of engineers. Dr. Wm. A. Wulf is president of the National Academy of Engineering.

The **Institute of Medicine** was established in 1970 by the National Academy of Sciences to secure the services of eminent members of appropriate professions in the examination of policy matters pertaining to the health of the public. The Institute acts under the responsibility given to the National Academy of Sciences by its congressional charter to be an adviser to the federal government and, upon its own initiative, to identify issues of medical care, research, and education. Dr. Harvey V. Fineberg is president of the Institute of Medicine.

The **National Research Council** was organized by the National Academy of Sciences in 1916 to associate the broad community of science and technology with the Academy's purposes of furthering knowledge and advising the federal government. Functioning in accordance with general policies determined by the Academy, the Council has become the principal operating agency of both the National Academy of Sciences and the National Academy of Engineering in providing services to the government, the public, and the scientific and engineering communities. The Council is administered jointly by both Academies and the Institute of Medicine. Dr. Ralph J. Cicerone and Dr. Wm. A. Wulf are chair and vice chair, respectively, of the National Research Council.

www.national-academies.org

*This report is dedicated to the memory
of our fellow committee member,
Dr. Herbert Rosenkranz*

Preface

The U.S. Environmental Protection Agency (EPA) has been given authority to regulate a variety of environmental agents that might harm public health or the environment. Toxicity testing in laboratory animals provides much of the information used by EPA to assess the hazards and risks associated with exposure to environmental agents. The number of regulations, initiatives, and directives that require toxicity testing is growing. Therefore, EPA recognized the need for a comprehensive review of established and emerging toxicity-testing methods and strategies and asked the National Research Council (NRC) to conduct such a review and to develop a long-range vision and strategy for toxicity testing.

In this report, the NRC's Committee on Toxicity Testing and Assessment of Environmental Agents reviews current toxicity-testing methods and selected aspects of several reports by EPA and others on the topic of toxicity testing and assessment. A second report will present the committee's long-range vision and strategic plan to advance the practices of toxicity testing and human health assessment of environmental contaminants.

This report has been reviewed in draft form by persons chosen for their diverse perspectives and technical expertise in accordance with procedures approved by the NRC's Report Review Committee. The purpose of this independent review is to provide candid and critical comments that will assist the institution in making its published report as sound as possible and to ensure that the report meets institutional standards of objectivity, evidence, and responsiveness to the study charge. The review comments and draft manuscript remain confidential to protect the integrity of the deliberative process. We wish to thank the following for their review of this report: George Daston (Proctor and Gamble), David Dorman (CIIT Centers for Health Research), Alan Goldberg (Johns Hopkins University), A. Wallace Hayes (Harvard University) George Maldonado (University of Minnesota), Roger McClellan (Albuquerque, New Mexico), John Quackenbush (Harvard University), Lorenz Rhomberg (Gradient Corporation), I. Glen Sipes (University of Arizona), and Errol Zeiger (Consultant).

Although the reviewers listed above have provided many constructive comments and suggestions, they were not asked to endorse the conclusions or recommendations, nor did they see the final draft of the report before its release. The review of this report was overseen by the review coordinator Donald Gardner (Inhalation Toxicology Associates), and the review monitor, Donald Mattison (National Institutes of Health). Appointed by the NRC, they were responsible for making certain that an independent examination of this report was carried out in accordance with institutional procedures and that all review comments were carefully considered. Responsibility for the final content of this report rests entirely with the committee and the institution.

The committee gratefully acknowledges the following for making presentations to the committee: Hugh Barton (Environmental Protection Agency), Alan Boobis (Imperial College of London), and James Lamb (Weinburg Group) who spoke on behalf of the Health and Environmental Sciences Institute's Technical Committee on Agricultural Chemical Safety Assessment; Nancy Doerrer and Michael Holsapple (Health and Environmental Sciences Institute); Felix Frueh, David Hattan, and John Leighton (Food and Drug Administration); Laura Gribaldo and Thomas Hartung (European Centre for the Validation of Alternative Methods); William Farland, Michael Firestone, Jim Jones, Robert Kavlock, Carol Kimmel, and Hal Zenick (Environmental Protection Agency); Christopher Portier (National Institute of Environmental Health Sciences); and Lorenz Rhomberg (Gradient Corporation).

The committee is also grateful for the assistance of the NRC staff in preparing this report. Staff members who contributed to this effort are Ellen Mantus, project director; Roberta Wedge, project director through January 2005; Joanne Zurlo, director of the Institute for Laboratory Animal Research; James Reisa, director of the Board on Environmental Studies and Toxicology; Jennifer Obernier, program officer; Norman Grossblatt, senior editor; Jennifer Roberts, postdoctoral research associate; Mirsada Karalic-Loncarevic, research associate; Jordan Crago, senior project assistant; and Lucy Fusco, senior project assistant.

I would especially like to thank all the members of the committee for their efforts throughout the development of this report.

Dan Krewski, *Chair*
Committee on Toxicity Testing and
Assessment of Environmental Agents

Abbreviations

ACD	acute contact dermatitis
ACS	American Cancer Society
ADJ	adjustment factor
ADME	absorption, distribution, metabolism, and excretion
AEGL	acute exposure guideline level
AhR	aryl hydrocarbon receptor
ARNT	AhR nuclear translocator
BBDR	biologically based dose response
BMC	benchmark concentration
BMD	benchmark dose
BT	Buehler test
CDC	Centers for Disease Control and Prevention
CFR	Code of Federal Regulations
CNS	central nervous system
CSA	chemical safety assessment
CYPs	cytochrome P450s
DAF	dosimetric adjustment factor
EC	European Commission
ECVAM	European Centre for the Validation of Alternative Methods
EDSP	Endocrine Disruptor Screening Program
EDSTAC	Endocrine Disruptor Screening and Testing Advisory Committee
EGP	Environmental Genome Project
ELISA	enzyme-linked immunosorbent assay
EPA	Environmental Protection Agency

EPHT	environmental public-health tracking
EU	European Union
FDA	Food and Drug Administration
FEV_1	forced expiratory volume in 1 second
FFDCA	Federal Food, Drug, and Cosmetics Act
FIFRA	Federal Insecticide, Fungicide, and Rodenticide Act
FQPA	Food Quality Protection Act
GD	gestational day
GFP	green fluorescent protein
GMPT	guinea pig maximization test
HESI	Health and Environmental Sciences Institute
HPV	high-production volume
HTS	high-throughput screening
HUPO	Human Proteome Organization
IARC	International Agency for Research on Cancer
IC_{50}	inhibition concentrations 50%
ICCVAM	Interagency Coordinating Committee on the Validation of Alternative Methods
ICH	International Conference on Harmonization
IgM	immunoglobulin M
ILSI	International Life Sciences Institute
IOM	Institute of Medicine
IPCS	International Programme on Chemical Safety
LC_{50}	lethal concentration 50%
LD_{50}	lethal dose 50%
LLNA	local lymph node assay
LOAEL	lowest-observed-adverse-effect level
MPCA	microbial pest-control agent
MTS	medium-throughput screening
NAAQS	national ambient air quality standards
NICEATM	NTP Interagency Center for the Evaluation of Alternative Toxicological Methods
NIEHS	National Institute of Environmental Health Sciences
NOAEL	no-observed-adverse-effect level
NRC	National Research Council
NTP	National Toxicology Program
OECD	Organisation for Economic Co-operation and Development
OSHA	Occupational Safety and Health Administration
PAH	polycyclic aromatic hydrocarbon

PBPK	physiologically based pharmacokinetic
PBT	persistence, bioaccumulation, and toxicity
PK	pharmacokinetic
PMN	premanufacturing notice
PND	postnatal day
QSAR	quantitative structure-activity relationship
REACH	Registration, Evaluation and Authorisation of Chemicals
RfC	Reference concentration
RfD	Reference dose
SAR	structure-activity relationship
SIDS	screening information dataset
SNP	single-nucleotide polymorphism
SRBCs	sheep red blood cells
TCDD	2,3,7,8-tetrachlorodibenzo-p-dioxin
TG	testing guideline
TGD	technical guidance document
TSC	the SNP Consortium
TSCA	Toxic Substances Control Act
UDP	up-down procedure
UF	uncertainty factor
UN	United Nations
vPvB	very persistent and very bioaccumulative
WHO	World Health Organization

Contents

FIGURES

TABLES

Toxicity Testing for Assessment of Environmental Agents

Interim Report

Summary

In the United States, several federal agencies have been given authority to regulate a variety of environmental agents that might harm public health. Accordingly, the agencies implement regulations that establish maximum acceptable concentrations of environmental agents in drinking water, set permissible limits of exposure of workers, define labeling requirements, establish tolerances for pesticides residues on food, and set other kinds of limits on the basis of risk assessment. Toxicity testing in laboratory animals provides many of the data needed for risk assessment, such as information on the possible effects of exposure to a substance and the exposure concentrations at which effects might be observed.

New directives and initiatives for toxicity testing in the United States and Europe reflect an increased demand for toxicity information to provide a rational basis for regulating environmental agents. At the same time, new testing technologies and methods have continued to emerge. The U.S. Environmental Protection Agency (EPA) recognized the need for a comprehensive review of established and newly developed toxicity-testing methods and strategies and asked the National Research Council (NRC) to conduct an independent review and to develop a long-range vision and strategy for toxicity testing. In response to EPA's request, the NRC convened the Committee on Toxicity Testing and Assessment of Environmental Agents.

1

COMMITTEE'S CHARGE AND APPROACH TO ITS CHARGE

The committee was asked to conduct a two-part study to assess and advance current approaches to toxicity testing and assessment undertaken to meet regulatory data needs. For the first part of the study, the committee was asked to review relevant aspects of several reports by EPA and others on the topic of toxicity testing and assessment. For the second part, the committee was asked to develop a long-range vision and strategic plan to advance the practices of toxicity testing and human health risk assessment of environmental contaminants. The present report fulfills the first part of the two-part study. The second report is expected to be completed by fall 2006. The committee was asked to focus on human toxicology and was not charged with reviewing toxicity testing and strategies designed to evaluate ecologic effects of environmental agents.

The committee heard presentations from representatives of several EPA offices, other federal agencies, and a number of organizations at public sessions, and it considered numerous documents and resources. The committee structured its review by first considering current toxicity-testing protocols. Recognizing that human data can be the most relevant for human health risk assessment, the committee considered the various types of human data available and the impediments that often prevent the use of epidemiologic data in regulatory risk assessment. Testing strategies used to rank, screen, or characterize substances were reviewed next. Various guidance documents that discuss the use of toxicity data for human health risk assessment were then considered. Finally, the committee reviewed some near-term improvements in toxicity-testing approaches proposed by others and some emerging technologies that may advance the field of toxicity testing.

Most of the documents reviewed by this committee describe initiatives or proposals that are still under development. Some have few details, and some were available to the committee only as drafts. Therefore, the committee focused on major themes rather than details, and it reviewed the documents primarily to compare various overall testing strategies and to evaluate the potential for the strategies to improve testing of environmental agents. The committee primarily examined toxicity-testing strategies rather than protocols for individual assays. Regarding documents that included an array of issues, the committee focused on the sections that dealt directly with toxicity testing and strategies and did not review sections that discussed risk-assessment approaches and policy issues, which were considered outside the scope of the committee's task.

COMMITTEE CONCLUSIONS

Toxicity Testing

The goals of toxicity testing are to identify possible adverse effects of exposure to environmental agents, to develop dose-response relationships that can elucidate the severity of effects associated with known exposures, and ultimately to predict the effects of exposure of human populations. Over the last several decades, scientists have developed consensus testing protocols, which have been designed to minimize variance and bias, to reduce false-positive and false-negative results, and to balance desired information with costs and resources. Some toxicity tests are designed to evaluate general toxicity resulting from exposures of various durations—acute, subchronic, and chronic—and others are designed to evaluate specific health effects, including reproductive and developmental toxicity, neurotoxicity, immunotoxicity, genetic toxicity, and carcinogenicity. Toxicity tests may also be distinguished by their objectives—to evaluate final outcomes of a specified exposure duration; to characterize the possible modes of action of such outcomes, which can depend on exposure route, concentration, and duration; to characterize dose-response relationships; or to identify a potential hazard, such as carcinogenicity from the results of a genotoxicity assay.

Testing Strategies

Testing strategies vary considerably, although they can often be described by three basic testing approaches: battery, tiered, or tailored. A battery is a specific set of toxicity tests applied to all chemicals in a group. Testing batteries are sometimes intended to provide the minimal dataset necessary for risk-based screening, regulation, or management. In tiered testing, the results of a specific set of toxicity tests and risk-management needs are used to guide decisions about the nature and extent of further testing. A substance is assigned to a category and then moves through a series of tests sequentially with the data from each test informing the next step in the process. In tailored testing, information on exposure, suspected adverse effects, and mechanism of action is used to determine the scope of tests to be conducted on a given chemical or class of chemicals. Characterizing an overall testing strategy as a battery, tiered, or tailored approach is often impossible because testing strategies are typically combinations of these three basic approaches.

The toxicity tests and strategies discussed in this report have evolved primarily as a means of characterizing potential human health hazards and dose-response relationships at least at high doses. The information produced is often judged to be sufficient for decision-making. For example, test results may provide reasonable assurance that a food additive or pesticide can be safely used as proposed. In contrast, if the difference between toxic doses and relevant human exposures is not large, further testing may be needed to refine the dose-response relationship at lower doses and to answer questions concerning the mechanism of action. Alternatively, regulatory action may be used to reduce human exposures.

Different testing strategies generally stem from legislative mandates or from differences in the practices of individual agencies or program offices. Thus, different strategies have developed to evaluate pesticides and food additives, to screen new industrial chemicals, and to investigate specific health effects, such as endocrine disruption. Different approaches can result in inconsistent testing strategies among agencies or categories of chemicals even if the ultimate regulatory goal is the same. The nature and extent of toxicity testing ideally should be guided by the regulatory risk-management decisions to be made and the assessments needed to support them.

Human Data

Human data usually are not a part of toxicity-testing strategies despite the importance of human responses to potentially toxic agents. Although animal toxicity studies provide relevant information on potential adverse health effects of exposure to an agent, interspecies differences can cause effects relevant to the human population to be missed. A famous example is thalidomide, to which rats are highly resistant but human fetuses are exquisitely sensitive. Studying the human population also provides an opportunity to evaluate the effects of the full variety of agents in the complex contexts of workplaces and daily lives. Clearly, no population data will be available on a chemical newly introduced to the marketplace. Population data will be available only on chemicals that have been in production for some time, perhaps several decades. Thus, differences in data availability on new versus existing chemicals should be considered in developing the role of human data in any toxicity-testing strategy.

Human data come primarily from epidemiologic studies, which investigate the relationship between exposure to a substance and potential health effects in a human population. Such studies have often been criticized because of methodologic limitations that make it difficult to draw clear associations between particular exposures and potential health effects. Components of epidemiologic studies that have posed problems include the assessment of exposure, which often involves only uncertain or indirect estimates of human exposure, and evaluation of exposure-effect relationships, particularly for chemicals for which there is an indeterminate and possibly long period between exposure and manifestation of effect. However, emerging technologies and approaches, such as biomonitoring and molecular and genetic epidemiology, may overcome some of the limitations and will be discussed in greater detail in the committee's second report.

Use of Data in Human Health Risk Assessment

Data from animal toxicity testing, human studies, and in vitro methods are used in human health risk assessment to identify potential hazards, to characterize effects at different exposure levels, to determine the probability of adverse effects of given human exposure scenarios, and ultimately to establish environmental standards and exposure guidance levels. Regulatory agencies have developed noncancer and cancer risk-assessment guidelines that provide comprehensive guidance on use and interpretation of relevant data to set exposure limits to protect public health. In general, the guidelines for assessing hazard and dose-response relationships have coevolved with scientific developments and laboratory capabilities. In some respects, the data being generated correlate well with guideline requirements. In other respects, there is a disconnect between the data needed for risk assessment and the data generated in the laboratory or field. Three examples are provided below.

Typical cancer guidelines require direct evidence of cancer in animals or humans to classify a chemical as having carcinogenic potential. When such data are not available, the chemical is classified as having, for example, "inadequate information to assess carcinogenic potential"; cancer risk is not estimated; and the chemical is generally treated as posing zero cancer risk. A system for using indirect evidence, such as structure-activity information and mechanistic data, could be developed to guide the assessment of chemicals that lack adequate cancer bioassay or epidemiologic data. Similarly, systems and guidance could be created for identifying a potential for neurotoxicity, developmental toxicity, and other kinds of toxicity on the basis of short-term tests and high-throughput approaches that use end points that are more specific to processes that are conserved across species.

For mutagenic carcinogens or carcinogens of unknown mechanism, estimating risk from animal data assumes that each individual faces the same risk of cancer at a given dose. A generic uncertainty factor is used in noncancer guidelines to adjust for variability among people. Testing strategies do not reflect a systematic approach for developing data to assess the variability of human responses to chemicals quantitatively. Such data would aid in understanding whether the current procedures for estimating cancer risk are conservative overall or may in some cases understate the risk for some segments of the population.

The generation of data for mode-of-action evaluations (with the exception of standard genotoxicity testing) and pharmacokinetic modeling is typically ad hoc. The data may be supplied by interested parties or otherwise available in the literature but are generally not required by the regulatory agencies. Although the guidelines may provide a loose framework for those approaches, they provide little specific guidance on data-generation issues. Optimizing further testing to improve the initial characterization of a particular chemical or class of chemicals can be highly context-dependent; however, a general framework and further guidance on developing a testing strategy to improve specific risk assessments would be useful.

Proposals to Improve Toxicity-Testing Strategies

The committee's review of current toxicity-testing strategies reveals a system that is reaching a turning point. Agencies typically have responded to scientific advances and emerging challenges by simply altering individual tests or adding tests to the existing regimens. That patchwork approach has not provided a fully satisfactory solution to the fundamental problem, which appears to be a tension among four objectives that are difficult to meet simultaneously:

- *Depth*, providing the most accurate, relevant information possible for hazard identification and dose-response assessment.
- *Breadth*, providing data on the broadest possible universe of chemicals, end points, and life stages.
- *Animal welfare*, causing the least animal suffering possible and using the fewest possible animals.
- *Conservation*, minimizing the expenditure of money and time on testing and regulatory review.

The committee acknowledges that meeting all four objectives poses a substantial challenge.

Several agencies or organizations have evaluated various toxicity-testing strategies with the goal of addressing gaps and inefficiencies in current approaches. The following sections highlight the committee's findings on proposals by EPA, the Health and Environmental Sciences Institute of the International Life Sciences Institute (ILSI-HESI), the European Union (EU), and the National Toxicology Program (NTP). More detailed discussion is provided in Chapter 6 of the committee's report.

EPA Review

In its 2002 report *A Review of the Reference Dose and Reference Concentration Processes*, EPA reviewed its procedures for deriving reference values and specifically the adequacy of the toxicity tests to accomplish that purpose. The committee focused its review on Chapter 3 of the EPA report because that chapter directly addressed toxicity-testing approaches. The committee did not critique the other chapters on risk-assessment approaches and application of uncertainty factors, which were considered outside the scope of the committee's task.

EPA's report raised five major issues: (1) the presence of data gaps in current toxicity-testing approaches, (2) a possible need to refine acute-toxicity testing protocols to support short-term risk assessments, (3) concerns about methods to incorporate pharmacokinetic and pharmaco-dynamic data into toxicity-testing approaches, (4) questions regarding incorporation of data on direct dermal toxicity into reference dose (RfD) development, and (5) a need to reconsider current toxicity-testing strategies systematically with an eye to improving efficiency and effectiveness.

First, the committee agrees that there are numerous data gaps in life stages and specific health effects evaluated in current toxicity-testing approaches. Few data are available to determine the degree to which those gaps have practical significance for risk assessment or whether they are primarily of theoretical or academic concern. The committee cautions against adding testing requirements only for the sake of theoretical thoroughness, because such an approach could result in substantial waste of animals and resources with little gain. However, the extent to which the data gaps might have practical consequences for risk assessment should be evaluated, and a reasonable interim approach to address this problem should be generated. Modest changes in existing protocols could enhance the array of health effects and life stages evaluated, and the resulting findings could trigger more in-depth testing of specific outcomes and life stages where it is warranted. The committee notes that epidemiologic studies with reliable exposure assessments could shed some light on the likelihood that current toxicity tests are missing important health effects or are not adequate for evaluating different life stages.

Second, the committee agrees that the existing protocols for acute toxicity testing focus on lethal effects and gross observations and generally do not provide adequate information for acute and short-term RfDs or reference concentrations (RfCs). Conducting acute protocols that address latency, reversibility, and differential susceptibility for all toxicity outcomes currently required in subchronic and chronic protocols would lead to very complex animal studies. Before such complex protocols are conducted, acute lethality studies, repeated-dose toxicity studies, and human data should be evaluated to determine the need for the more complex studies and ultimately to guide the design of these studies.

Third, the committee agrees that generally little information is available on pharmacokinetics, including possible differences across life stages. It is critically important to define the purpose of pharmacokinetic studies to avoid the creation of data that are unlikely to be used and therefore represent a waste of animals, time, and resources. Additional data should not be routinely required, but the need should be evaluated case by case.

Fourth, the committee finds that the relevant exposure route and exposure durations should be considered in developing a testing strategy. When dermal exposure is a primary exposure route, there is a general need for better data on dermal uptake and absorption. However, it is important to consider whether skin is an important route of exposure before beginning the process of setting a dermal RfD. Worker data and clinical reports could be collected more systematically and used preferentially in setting dermal reference doses of existing chemicals.

Finally, the committee agrees that a new strategy is needed to improve efficiency, reduce animal use, increase the number of chemicals screened for toxicity, and address some of the data gaps identified. EPA explored alternative testing protocols for acute and chronic toxicity testing to stimulate new ideas. It did not articulate how such protocols might be incorporated into a testing strategy. The committee supports the notion of expanded tests that combine studies to conserve resources and provide more in-depth evaluations of outcomes and life stages. However, considerable development and evaluation may be required to ensure that tests are feasible and reproducible, do not compromise study sensitivity, produce the desired data, and reduce the use of animals. Expanded bioassays may ultimately have a role in selectively testing high-priority chemicals but might not necessarily be amenable to widespread application.

ILSI-HESI Draft Proposals

The committee reviewed a testing strategy proposed by ILSI-HESI and various recommendations contained in its draft reports: *Systemic Toxicity White Paper*; *Life Stages White Paper*; and *The Acquisition and Application of Absorption, Distribution, Metabolism, and Excretion (ADME) Data in Agricultural Chemical Safety Assessments*. ILSI-HESI proposed substantive modifications of toxicity-testing requirements for pesticides and identified some potential omissions and redundancies in current pesticide testing. Recommendations included changing exposure durations of required toxicity tests, eliminating some required guideline studies, modifying some studies to enhance evaluation of specific health effects, and generating chemical-specific pharmacokinetic data to inform study design and data interpretation.

The committee supports the general approach used by ILSI-HESI to tailor testing to meet risk-assessment needs. Specifically, ILSI-HESI proposed using exposure considerations (such as the difference between doses that produce effects in animals and expected human exposure to pesticides) to provide a conceptual framework for guiding the selection and extent of testing. That approach, however, may not be useful for chemicals for which the degree and circumstances of human exposure are difficult to predict.

The committee supports the general ILSI-HESI approach of using existing databases to evaluate the importance of specific toxicity tests or their contribution to the dataset and endorses further broad retrospective reviews. However, the committee has concerns about the recommended elimination of some toxicity tests from first-tier testing. For example, ILSI-HESI proposed removing the rat teratology study and using an extended one-generation study and a rabbit teratology study to evaluate developmental effects. Although the proposed one-generation study substantially improves postnatal evaluation of many nonreproductive outcomes, it is unclear whether it would be as sensitive as a rat teratology study for prenatal developmental-toxicity outcomes or would adequately reveal the potential hazard and trigger a followup study. Furthermore, EPA often bases acute reference values on the rat teratology study. In contrast, postnatal effects in a one-generation study are not typically used for deriving acute reference values. The effect of eliminating the rat teratology study on hazard identification and on the setting of acute reference values should be evaluated if the proposal is pursued.

Overall, the changes proposed by ILSI-HESI may affect the probability of finding some effects and change the volume of evidence available to an assessor in judging the presence or importance of an effect. Cumulatively, it is unclear how the different aspects of the proposal would affect the overall fidelity of the testing process. The committee notes that the ILSI-HESI evaluation may have overlooked redundancy of testing as a critical part of the weight-of-evidence approach. More-limited testing and less redundancy could mean less confirmatory evidence and greater potential overall for reduced sensitivity of the testing strategy. Making decision-making more conservative, erring in the direction of false-positive results, or using greater uncertainty factors may address those issues. Corresponding adjustments of risk-assessment guidelines that emphasize positive results of multiple studies for confirmatory evidence also may address those issues.

REACH Program

The EU is engaged in a bold effort to restructure its approach to toxicity testing. The primary goal of the new approach, known as REACH (Registration, Evaluation and Authorisation of Chemicals), is outlined in the 2004 EU report *The REACH Proposal Process Description*. The goal is to collect data on and regulate about 30,000 chemicals produced or imported in excess of 1 metric ton per year on which there are limited toxicity and environmental data. The new approach is based on production or importation volume, which serves as a surrogate of potential human exposure. It specifies a battery of tests or specific effects to be evaluated at each level without being prescriptive about how tests will be done. The committee notes that the approach enhances flexibility but may make comparison of results difficult. Also, although tonnage may be an initial rough surrogate of potential human exposure, other information (such as whether the chemical is an intermediate to which humans are unlikely to be exposed) may also be relevant.

The committee found that the REACH program focuses more on screening large numbers of chemicals than on generating in-depth information that is often needed for quantitative risk assessment. However, the REACH program does allow for greater depth of testing to be triggered on the basis of initial results. The REACH program has the advantage of generating at least some toxicity data on chemicals that are not now subject to testing in the United States.

NTP Roadmap for the Future

In its 2004 report *The NTP Vision for the 21st Century*, the NTP discussed its goals: to refine traditional toxicity assays; to develop rapid, mechanism-based predictive screens for environmentally induced diseases; and to improve the overall use of NTP toxicity-testing assays for public-health decisions. The NTP also described its current research initiatives:

- To review and refine toxicity-testing protocols.
- To incorporate new approaches, such as genomic analyses, into toxicity-testing strategies.
- To improve the use of pharmacokinetic information in toxicologic evaluation.
- To explore the use of nonmammalian alternatives to toxicity testing.
- To expand the use of imaging technologies for detecting and quantifying molecular and cellular lesions and for improving the speed and precision of pathology reviews.

The committee found that the NTP's near-term efforts to refine and extend its toxicity tests and to improve the use of pharmacokinetic information promise to increase the depth of toxicity information on chemicals assayed and to provide greater insight in applying the findings to humans. However, as acknowledged by the NTP, the resulting portfolio would still be resource-intensive and incapable of addressing large numbers of chemicals that require some level of toxicity assessment. That problem emphasizes the importance of the NTP's long-term goal to develop screening strategies that use nonanimal models. Such a focus by an agency like the NTP is needed if those approaches are to become viable alternatives to traditional toxicity testing in animals.

Future Directions

The committee identified several recurring themes and questions in the various reports that it was asked to review. The recurring themes included the following:

- The inherent tension between breadth, depth, animal welfare, and cost of toxicity testing and the challenge to address any one of these issues without worsening another.
- The importance of distinguishing between testing protocols and testing strategies as one considers modifications of current testing practices.
- The need to be cautious in adding testing requirements for the sake of theoretical thoroughness.
- The possible dangers in making tests so efficient, such as by eliminating all overlap, that there are no means to verify results.
- The role of both uniform testing protocols and strategies to enhance comparability and chemical-specific tailored testing in deepening understanding of a particular chemical's mode of action.
- The importance of recognizing that toxicity testing for regulatory purposes should be conducted primarily to serve the needs of risk management.

The recurring questions that arose during the committee's review and its initial observations are provided below. The questions and observations will help to frame the discussion for the committee's second report, which will provide a long-range vision and strategic plan for advancing the practices of toxicity testing and human health risk assessment.

Which environmental agents should be tested? All new and existing environmental agents should be evaluated; however, the intensity and depth of testing should be based on practical needs, including the use of the chemical, the likelihood of human exposure, and the scientific questions that such testing must answer to support a reasonable science-policy decision. Fundamentally, the design and scope of a toxicity-testing approach need to reflect risk-management needs.

How should priorities for testing chemicals be set? Priority-setting should be a key component of any testing strategy that is designed to address a large number of chemicals, and a well-designed scheme is essential for systematic testing of industrial chemicals on which there are few data. It makes sense to consider exposure potential in designing test strategies. Chemicals to which people are more likely to be exposed or to which some populations may receive relatively high exposures—whether they are pesticides or industrial chemicals—should undergo more in-depth testing. This concept is embedded in several existing and proposed strategies. In some strategies, production volume is the primary measure of potential human exposure; but production volume alone may not be the best surrogate of human exposure. Other important factors to consider are use, exposure patterns, and a chemical's environmental persistence and bioaccumulation, which is important because of the potential for increasing exposure over time and continuing exposure even after use has ceased.

What strategies for toxicity testing are the most useful and effective? Current approaches to toxicity testing include testing batteries, tiered testing, tailored testing, and a combination of the three. The committee finds that there are pros and cons of various approaches but leans toward tiered testing with the goal of focusing resources on the evaluation of the more sensitive adverse effects of exposures of greatest concern rather than full characterization of all adverse effects irrespective of relevance for risk-assessment needs. The committee, however, notes that tiered-testing approaches should be designed to expedite regulatory decisions and to discourage toxicity testing that is not used to address regulatory questions.

How can toxicity testing generate data that are more useful for human health risk assessment? Many have criticized existing approaches to toxicity testing on the grounds that the data generated are often not ideal for conducting human health risk assessment. Extrapolations are often made with weak scientific justifications, and uncertainty factors are used to bridge the gaps. The current proposals to improve toxicity-testing strategies, discussed above, are unlikely to solve the fundamental problem. The committee cautions against indiscriminately generating large amounts of data with an eye to creating optimal datasets for characterizing risks posed by single chemicals. Emerging technologies and approaches, such as "-omics" technologies and computational toxicology, may help to address the problem.

How can toxicity testing be applied to a broader universe of chemicals, life stages, and health effects? There are major gaps in current toxicity-testing approaches. The importance of the gaps is a matter of debate and depends on whether effects of public-health importance are being missed by current approaches. However, it is impractical to test every chemical for every possible health effect over all life stages. The emphasis should be on chemicals that have the greatest potential for human exposure. The emerging technologies may help to screen chemicals more rapidly and to indicate a need for further testing.

How can environmental agents be screened with minimal use of animals and efficient expenditure of time and other resources? One strategy that can be applied to reduce animal use is the grouping of chemicals of similar structural class and the in-depth testing of only one or a few representative chemicals; risk assessments of all chemicals in the class would be based on the resulting data. In grouping chemicals, known modes of action should be emphasized. Such strategies should address any data needed to support application of study findings to other chemicals in the group. Newer approaches also have great promise.

How should tests and testing strategies be evaluated? Testing strategies may be evaluated in terms of the value of information they provide in light of the four objectives—increasing depth of knowledge for more accurate risk assessment; increasing coverage of chemicals, life stages, and end points; preserving animal welfare; and minimizing cost. In evaluating new tests and testing strategies, there remains the difficult question of what is to serve as a "gold standard" for performance. Simply comparing the outcomes of new tests with the outcomes of current tests may not be the best approach; whether it is will depend on the reliability and relevance of the current tests. Ideally, regulations and risk-assessment guidelines will evolve with testing capabilities and scientific understanding. That issue will increase in importance with greater use of screening approaches (for example, in vitro tests, gene arrays, and mode-of-action screens) that produce indirect evidence on both cancer and noncancer end points.

1

Introduction

The purpose of toxicity testing is to generate information about a substance's toxic properties so that the health and environmental risks it poses can be adequately evaluated. Federal agencies use information from toxicity testing to establish acceptable concentrations of environmental agents in drinking water, to set permissible exposure limits for workers, to establish tolerances for pesticide residues on food, to register and re-register pesticides, and ultimately to protect public health and the environment. As reflected in new directives and initiatives for toxicity testing in the United States and Europe, the demand for toxicity information to provide a rational basis for regulating environmental agents has increased. At the same time, testing technologies and methods have continued to emerge. Thus, the U.S. Environmental Protection Agency (EPA) recognized the need for a comprehensive review of established and emerging toxicity-testing methods and strategies and asked the National Research Council (NRC) to conduct such a review and to develop a long-range vision and strategy for toxicity testing. In response to EPA's request, the NRC convened the Committee on Toxicity Testing and Assessment of Environmental Agents, which prepared this report.

REGULATORY REQUIREMENTS

The U.S. Congress has enacted laws calling for limits on chemical exposures that "provide an ample margin of safety to protect public health" (Clean Air Act; 42 USC §7412(f) [2003]), "assure protection of public health" (Clean Water Act; 33 USC §1312(a) [2003]), provide "a reasonable certainty that no harm will result" (Food Quality Protection Act; 21 USC §346a(b) [2003]), and "adequately assures, to the extent feasible, on the basis of the best available evidence, that no employee will suffer material impairment of health or functional capacity" (Occupational Safety and Health Act; 29 USC §655(b) [2003]) (see Table 1-1). Federal agencies implement those statutes by promulgating standards or adopting guidance levels—such as air-quality criteria, maximum contaminant levels for drinking water, pesticide-residue tolerances, and permissible exposure limits for workplaces—that limit people's exposure to chemicals. The standards and guidance levels are often developed through human health risk assessment, although other factors—such as treatment technology, feasibility, benefits, and costs—may also be considered. Toxicity testing in laboratory animals provides much of the information needed to characterize the nature and extent of the risk so that appropriate risk-management action can be taken.

TOXICITY TESTING

Many factors are determinants of health, including socioeconomic status, birth weight, sex, genetics, diet, pathogens, smoking habits, cultural activities, and the environment. Exposures to environmental agents contribute to the aggregate effects of the other factors, but the nature and magnitude of the contribution are often debated. The results of toxicity testing can help to clarify risks to health posed by environmental exposures and provide support for effective risk-management decisions.

Toxicity-testing requirements to evaluate effects on human health often involve studies of whole animals, typically rats, mice, dogs, and rabbits, although other species, including humans, can be used. Exposures can range from short-term (for example, an hour) to long-term (for example, 2 years) and be continuous or episodic or consist of a single event. Tests may focus on a particular life stage, sex, or condition of exposure. The effects evaluated can be numerous and can include such diverse outcomes as subtle behavioral changes, impairment of reproduc-

TABLE 1-1 Some Major U.S. Regulatory Agency Authority, Public-Health Goals, and Risk-Management Approaches

Agency and Program	Enabling Statute and Citation	Statutory Mandate: Qualitative Health and Safety Goal	Implementation: Regulatory Risk-Management Requirement or Advisory
EPA, water	Safe Drinking Water Act/42 USC § 300g-1 (2003)	• Establish contaminant concentrations at which "no known or anticipated adverse effects on the health of persons occur and which allows an adequate margin of safety"; may consider feasibility	• Maximum contaminant levels • Maximum contaminant level goals
	Federal Water Pollution Control Act (Clean Water Act)/33 USC §§ 1312-1333 (2003)	• Control pollutant discharges to "assure protection of public health"; may consider feasibility • Establish criteria that "protect public health and welfare"	• Technology-based effluent limitations • Water-quality standards
EPA, pesticides	Food Quality Protection Act[a]/21 USC § 346a(b) (2003)	• Ensure that there is "a reasonable certainty that no harm will result from aggregate exposure to pesticide chemical residues"; considers benefits	• Pesticide-residue tolerances • Additional margin of safety to protect infants and children
	Federal Insecticide, Fungicide, and Rodenticide Act/7 USC §§ 136a, 136w (2003)	• Prevent "unreasonable adverse effects on the environment" (includes humans); considers benefits • Protect children and adults from serious injury or illness	• Registration and use limits • Safe-packaging requirements

(Continued)

TABLE 1-1 Continued

Agency and Program	Enabling Statute and Citation	Statutory Mandate: Qualitative Health and Safety Goal	Implementation: Regulatory Risk-Management Requirement or Advisory
EPA, air	Clean Air Act/42 USC §§ 7408(a), 7412(f) (2003)	• Avoid emissions that "may reasonably be anticipated to endanger public health or welfare" • Promulgate standards that "provide an ample margin of safety to protect public health"	• Air-quality criteria for pollutants • Technology-based standards to limit emissions of hazardous air pollutants ("maximum achievable control technology") • Further regulation needed if residual risk exceeds 10^{-4} or is of public-health significance
EPA, toxic substances	Toxic Substances Control Act/15 USC § 2603 (2003)	• Determine that a substance "does not or will not present unreasonable risk of injury to health"	• Toxicity testing when information available is insufficient to determine or predict health effects
EPA, Superfund	Superfund Amendments and Reauthorization Act[b]/42 USC § 9621 (2003)	• Establish site-cleanup standards that are "protective of human health"; may consider feasibility	• Remedial action
OSHA	Occupational Safety and Health Act/29 USC § 655(b) (2003)	• "Set the standard which most adequately assures, to the extent feasible, on the basis of the best available evidence, that no employee will suffer material impairment of health or functional capacity"	• Occupational safety or health standards • Permissible exposure limits

FDA, food safety	Federal Food, Drug, & Cosmetic Act[c]/21 USC §§ 348(a) (2003), 379e (2001)	• Determine that the intended use of a food additive, food-contact substance, or color additive is "safe" • No human or animal carcinogens	• Regulation providing conditions of safe use • Requirement of notification for food-contact substances and petition for food additive • Listing for use as a color additive • Classification as generally recognized as safe
FDA, drug evaluation	21 USC § 355(d) (2001)	• Show whether a drug is "safe for use," as well as efficacious	• Approval of new drug applications, considering risks and benefits
CPSC	Consumer Product Safety Act/15 USC §§ 2057, 2058 (2003)	• "Eliminate or reduce an unreasonable risk of injury associated with" consumer products; considers costs and benefits	• Mandatory safety standards • Bans • Recalls

[a] Amends the Federal Food, Drug, and Cosmetic Act.
[b] Reauthorizes the Comprehensive Environmental Response, Compensation, and Liability Act (Superfund).
[c] As amended by the Food and Drug Modernization Act.
Abbreviations: CPSC, Consumer Product Safety Commission; EPA, U.S. Environmental Protection Agency; FDA, Food and Drug Administration; OSHA, Occupational Safety and Health Administration.

tion, abnormal development, alterations in gene function, organ toxicity, cancer, and death. New methods that rely on molecular biology, information technology, and alternatives to whole-animal testing are emerging and may provide information that allows better extrapolation of results in test species to the genetically diverse human population. Some new methods may eventually replace various traditional toxicity tests.

Federal agencies and international organizations—including EPA, the National Toxicology Program (NTP) (Chhabra et al. 1990), the Food and Drug Administration (FDA 1997, 2004), and the Organisation for Economic Co-operation and Development (OECD 2004, 2005)—have developed documents that provide guidance on testing protocols. Testing guidelines are intended to convey to members of the regulated community what is expected of them and provide a uniform and sometimes flexible approach to toxicity testing that produces comparable results. The OECD protocols serve not only as standards but as means to harmonize requirements among regulatory authorities and thus reduce repetition of studies. The harmonization efforts should increase efficiency and reduce animal use.

CHALLENGES TO TOXICITY TESTING AND ASSESSMENT

The continuing challenge is to determine the best methods for extrapolating from the exposure conditions and effects observed in the laboratory to those relevant to the human environment. Toxicity-testing data can be used in various ways to aid in that extrapolation. Pharmacokinetic data can provide a better understanding of the qualitative and quantitative comparability of the relationship between exposure and dose in test species and in humans. Population-based studies that examine effects on exposed humans can provide information that improves extrapolation from laboratory-animal data to humans or in some cases eliminates the need to rely on laboratory-animal data altogether. Studies that provide a quantitative understanding of the difference in susceptibility to a chemical between humans and test species can be used to develop an interspecies adjustment factor based on scientific data rather than science policy. Finally, studies that provide an understanding of variations in susceptibility to the effects of a substance in different populations or life stages can help to identify substances that require special regulatory attention to protect sensitive groups and may also identify exposures that will have no deleterious effects even in sensitive people.

Studies like those discussed have resulted in much-improved human risk estimation, but substantial challenges remain. Most toxicity tests use exposures that exceed environmental exposures by several orders of magnitude to improve test sensitivity, but high exposures can distort the specificity of a test and its qualitative and quantitative applicability to actual human exposure. Scientific developments and new test methods are needed to address people's multiple simultaneous chemical exposures, their potential interactions, and the many factors that affect people's susceptibility to chemical exposures. Thus, even extensive testing and an accurate understanding of biologic modes of action cannot predict exactly what will happen in a diverse human population under environmental conditions of exposure. Precise descriptions of risk are desirable to protect public health, but they remain elusive. Nevertheless, even with these challenges and uncertainties, toxicity-testing data provide critical information for assessing hazard and risk potential and will continue to play a critical role in rational decision-making.

As noted above, EPA and other federal agencies have statutory responsibilities for obtaining and evaluating animal and human toxicity data for regulatory decision-making purposes. The numbers of health outcomes and questions that must be considered have also grown over the years. EPA and others have responded to the increased need to address various outcomes by developing risk-assessment guidelines and testing requirements, such as the risk-assessment guidelines for neurotoxicity (IPCS 2001; OECD 2004; EPA 1998), guidelines on children's cancer risk (EPA 2005), testing guidelines for developmental neurotoxicity (OECD 2004; EPA 1998) and guidelines for the use of genomics data (EPA 2004). Still, there is a growing recognition that because traditional toxicity testing approaches are time consuming and resource intensive, a large volume of existing and newly introduced chemicals cannot be adequately assessed using current testing practices. EPA recognized the need to review traditional toxicity testing approaches, new data-generation methods, and testing strategies comprehensively and asked NRC to perform such a review.

COMMITTEE'S TASK AND APPROACH

The committee members were selected for their expertise in developmental toxicology, reproductive toxicology, neurotoxicology, immunology, pediatrics and neonatology, epidemiology, biostatistics, in vitro methods and models, molecular biology, pharmacology, physiologically based pharmacokinetic and pharmacodynamic models, genetics, toxicogenomics, cancer hazard assessment, and risk assessment.

The committee was asked to conduct a two-part study to assess and advance current approaches to toxicity testing and assessment undertaken to meet regulatory data needs. For the first part of the study, the committee was asked to review selected aspects of several relevant reports by EPA and others. Those reports included EPA's 2002 review of the reference-dose and reference-concentration processes (EPA 2002), the International Life Sciences Institute (ILSI) Health and Environmental Sciences Institute (HESI) work to develop a tiered toxicity-testing approach for agricultural-chemical safety evaluations (ILSI-HESI 2004a,b,c), the work of the Interagency Coordinating Committee on the Validation of Alternative Methods to develop and validate alternatives to animal testing (ICCVAM 1997), pertinent NRC reports, and current work of NRC standing committees. Those reports were to be evaluated for specific elements, including analysis of current and anticipated regulatory needs, discussion of the current and planned inventory of toxicity-testing and assessment schemes and methods, evaluation of potential uses and limitations of new or alternative testing methods and analysis of how they might influence or define future testing strategies, and discussion of scientific advances that could affect the nature of information needed to assess potential human toxicity more completely. The present report was prepared to fulfill the first part of the study.

For the second part of the study, the committee was asked to build on the work presented in this report and develop a long-range vision and strategic plan to advance the practices of toxicity testing and human health assessment of environmental contaminants. The second report is expected to be completed by fall 2006.

To accomplish the task of preparing its first report, the committee held four meetings. The first three, held from June 2004 to November 2004, included public sessions. At the public sessions, the committee heard presentations from staff of several EPA offices, including representatives from the Office of Research and Development, the Office of Pesticide Programs, the Office of Children's Health, the National Health and Environmental Effects Research Laboratory, and the National Center for Environmental Assessment. The committee also heard presentations from staff of other federal agencies and organizations, including the NTP, the FDA Center for Drug Evaluation and Research and Center for Food Safety and Applied Nutrition, the European Center for the Validation of Alternative Methods, and the ILSI HESI and Risk Sciences Institute. The committee considered numerous documents, including those mentioned in the statement of task but also others, such as FDA guidance on pharmacogenomic-data submissions (FDA 2005) and the NTP Roadmap for the Future (NTP 2004).

ORGANIZATION OF THE REPORT

In Chapter 2, the committee presents an overview of consensus-study protocols focusing primarily on EPA guidelines. The objective of that chapter is not to detail each type of testing protocol but rather to indicate the general types of whole-animal and in vitro toxicity testing now in use. Chapter 3 considers a variety of human-based studies ranging from clinical trials to epidemiologic studies. Challenges that have often prevented the use of epidemiologic studies in regulatory risk assessment and possible advances and improvements in epidemiology are discussed. Chapter 4 examines applications of toxicity tests in testing strategies that are used to rank, screen, or characterize chemical toxicity. Several examples of testing strategies are presented. The strategies identified are not meant to be exhaustive but to illustrate the array of toxicity tests that may be required under different circumstances. Chapter 5 provides an overview of risk-assessment guideline documents that deal with the use of toxicity data in human health risk assessment and concludes with observations regarding strengths and weaknesses of the current system for generating toxicologic data to assess environmental risks. Chapter 6 is the committee's assessment of the various, and often conflicting, demands on the regulatory toxicity-testing framework and a review of near-term and long-term approaches that hold promise for improving toxicity testing. The chapter includes comments on the portion of the EPA review of its reference-dose and reference-concentration process that is relevant to toxicity testing and comments on the proposed ILSI-HESI approaches for pesticides, the NTP Roadmap for the Future, and the European Union's program. Chapter 7 discusses alternatives to animal testing and a few emerging technologies, such as -omics approaches and computational toxicology. It concludes with a discussion of validation to emphasize the importance of evaluating new toxicity-testing methods to ensure that the information obtained from them is at least as good as, if not better than, conventional mammalian models.

REFERENCES

Chhabra, R.S., J.E. Huff, B.S. Schwetz, and J. Selkirk. 1990. An overview of prechronic and chronic toxicity/carcinogenicity experimental study designs and criteria used by the National Toxicology Program. Environ. Health Perspect. 86:313-321.

EPA (U.S. Environmental Protection Agency). 1998. Guidelines for Neurotoxicity Risk Assessment. EPA/630/R-95/001F. U.S. Environmental Protection Agency, Washington, DC [online]. Available: http://www.epa.gov/ncea/raf/pdfs/neurotox.pdf [accessed July 13, 2005].

EPA (U.S. Environmental Protection Agency). 2002. A Review of the Reference Dose and Reference Concentration Processes. Final Report. EPA/630/P-02/002F. Risk Assessment Forum, U.S. Environmental Protection Agency, Washington, DC [online]. Available: http://www.epa.gov/iris/RFD_FINAL%5B1%5D.pdf [accessed March 11, 2005].

EPA (U.S. Environmental Protection Agency). 2004. Guidance on Use of Genomics Data. Computational Toxicology Program, Office of Research and Development, Washington, DC [online]. Available: http://www.epa.gov/comptox/comptoxfactsheet.html#guidance [accessed April 21, 2005].

EPA (U.S. Environmental Protection Agency). 2005. Supplemental Guidance for Assessing Susceptibility from Early-Life Exposure to Carcinogens. EPA 630/R-03/003F. Risk Assessment Forum, U.S. Environmental Protection Agency, Washington, DC [online]. Available: http://www.epa.gov/iris/children032505.pdf [accessed April 22, 2005].

FDA (Food and Drug Administration). 1997. Guidance for Industry: S1B Testing for Carcinogenicity of Pharmaceuticals. U.S. Department of Health and Human Services, Food and Drug Administration, Center for Drug Evaluation and Research (CDER), Center for Biologics Evaluation and Research (CBER), Washington, DC [online]. Available: http://www.fda.gov/cder/guidance/1854fnl.pdf [accessed March 9, 1005].

FDA (Food and Drug Administration). 2004. Guidance for Industry: Studies to Evaluate the Safety of Residues of Veterinary Drugs in Human Food: Carcinogenicity Testing VICH GL28. Final Guidance. U.S. Department of Health and Human Services, Food and Drug Administration, Center for Veterinary Medicine, May 24, 2004 [online]. Available: http://www.fda.gov/cvm/guidance/guide141.pdf [accessed March 9, 2005].

FDA (Food and Drug Administration). 2005. Guidance for Industry. Pharmacogenomic data submissions. U.S. Department of Health and Human Services, Food and Drug Administration, Center for Drug Evaluation and Research (CDER), Center for Biologics Evaluation and Research (CBER), Washington, DC [online]. Available: http://www.fda.gov/OHRMS/DOCKETS/98fr/2003d-0497-gdl0002.pdf.

ICCVAM (Interagency Coordinating Committee on the Validation of Alternative Methods). 1997. Validation and Regulatory Acceptance of Toxicological Test Methods, A Report of the ad hoc Interagency Coordinating Committee on the Validation of Alternative Methods. NIH Publication No. 97-3981. National Institute of Environmental Health Sciences, Research Triangle Park, NC. March 1997 [online]. Available: http://iccvam.niehs.nih.gov/docs/guidelines/validate.pdf [accessed March 11, 2005].

ILSI HESI (International Life Sciences Institute Health and Environmental Sciences Institute). 2004a. Systemic Toxicity White Paper. Systemic Toxicity Task Force, Technical Committee on Agricultural Chemical Safety Assessment, ILSI Health Sciences Institute, Washington, DC. November 2, 2004.

ILSI HESI (International Life Sciences Institute Health and Environmental Sciences Institute). 2004b. Life Stages White Paper. Life Stages Task Force, Technical Committee on Agricultural Chemical Safety Assessment, ILSI Health Sciences Institute, Washington, DC. November 2, 2004.

ILSI HESI (International Life Sciences Institute Health and Environmental Sciences Institute). 2004c. The Acquisition and Application of Absorption, Distribution, Metabolism, and Excretion (ADME) Data in Agricultural Chemical Safety Assessments. ADME Task Force, Technical Committee on Agricultural Chemical Safety Assessment, ILSI Health Sciences Institute, Washington, DC. November 2, 2004.

IPCS (International Programme on Chemical Safety). 2001. Neurotoxicity Risk Assessment for Human Health: Principles and Approaches. Environmental Health Criteria 223. Geneva: World Health Organization [online]. Available: http://www.inchem.org/documents/ehc/ehc/ehc223.htm [accessed March 2, 2005].

NTP (National Toxicology Program). 2004. The NTP Vision for the 21st Century. National Toxicology Program, National Institute for Environmental Health, Research Triangle Park, NC [online]. Available: http://ntp-server.niehs.nih.gov/ntp/main_pages/NTPVision.pdf [accessed March 11, 2005].

OECD (Organisation for Economic Cooperation and Development). 2004. Guideline Document for Neurotoxicity Testing. ENV/JM/MONO(2004) 25. OECD Series on Testing and Assessment No. 20. Organisation for Economic Cooperation and Development, Paris [online]. Available: http://appli1.oecd.org/olis/2004doc.nsf/linkto/env-jm-mono(2004)25 [accessed April 1, 2005].

OECD (Organisation for Economic Cooperation and Development). 2005. Chemicals Testing: OECD Guidelines for the Testing of Chemicals - Sections 4: Health Effects [online]. Available: http://www.oecd.org/document/55/0,2340,en_2649_34377_2349687_1_1_1_1,00.html [accessed March 14, 2005].

2

Animal and In Vitro Toxicity Testing

Animals have been used as sentinels for early detection of potential risk to humans or as models to study the causes, pathogenesis, progression, and treatment of diseases. The latter use gave rise to the field of investigative toxicology, wherein animals are used as surrogates to predict possible adverse health effects to humans arising from chemical exposures. That approach is challenged by some people for scientific, ethical, and philosophic reasons, but the use of animal models to assess hazards and risks to humans continues to be the standard for protecting human health. Over the last several decades, scientists have developed standardized protocols for testing potentially hazardous chemicals to ensure sound scientific methods and generation of high-quality data that are critical for assessing human hazards and risks.

Toxicity testing in animals is conducted to identify possible adverse effects resulting from exposure to an agent and to develop dose-response relationships that allow evaluation of responses at other exposures. Toxicity tests are designed to minimize variance, bias, and the potential for false-positive and false-negative results. Those goals, however, are weighed in light of constraints on costs and other resources. The types and extent of human exposure are important considerations in designing toxicity studies for human health risk assessment. An understanding of duration, frequency, intensity, and routes of exposure and an understanding of chemical stability and possible chemical breakdown products are helpful in guiding the selection of the dosing regimen, the test medium, and the test material.

Animal toxicity studies conducted for regulatory submission typically are conducted in rats, mice, rabbits, and dogs with greater focus on rats. Testing guidelines generally require that common laboratory strains be used. At least three dose groups and a control group usually are required. For most toxicity tests, the U.S. Environmental Protection Agency (EPA) requires that the highest dose elicit signs of toxicity without compromising survival. EPA strongly recommends that the lowest dose not produce any evidence of toxicity. The numbers of animals required are defined in each study protocol and range from five rats per sex per dose in 28-day toxicity studies to 10 rats per sex per dose in subchronic studies to 50 rats per sex per dose in carcinogenicity assays. For developmental and reproductive studies, the litter is considered the experimental unit, and at least 20 litters per dose are required. The statistical power of a study is determined by the number of animals used and the sensitivity of the end point being evaluated.

This chapter provides an overview of consensus-study protocols developed or codified by several organizations, including EPA and the Organisation for Economic Co-operation and Development (OECD). EPA specifies the types and extent of toxicity data that it requires to make regulatory decisions regarding the risks and benefits associated with pesticide products in accordance with the Federal Insecticide, Fungicide, and Rodenticide Act (FIFRA) and the Federal Food, Drug, and Cosmetic Act (FFDCA). The specific data requirements are listed in the *Code of Federal Regulations* (CFR), Title 40, Subpart E, Part 158 (40CFR158). EPA also requires testing of industrial chemicals under the Toxic Substances Control Act (TSCA). EPA has harmonized the testing protocols that may be used in support of FIFRA registrations and TSCA test rules and has harmonized the guidelines with those of OECD. Appendix B of this report provides a list of EPA's harmonized health-effects test guidelines.

OECD also develops test guidelines and guidance documents to help to characterize potential hazards associated with new and existing chemicals. The OECD document, *Guidelines for the Testing of Chemicals* (OECD *Guidelines*), is a collection of the most relevant internationally agreed-on testing methods used by government, industry, and independent laboratories (OECD 2004a). OECD publishes the guidelines to relieve some of the burden of chemical testing and assessment in multiple countries. Appendix B provides a list of OECD's health-effects test guidelines.

In addition to its guidelines, OECD publishes a monograph series called guidance documents and detailed review documents that provide information on available testing methods and on how to use OECD guidelines in a testing strategy for classification of specific end points.[1] They also discuss when such testing is useful or necessary, end points of concern, approaches for statistical analysis, and limitations of tests. The detailed review documents are prepared when it is necessary to assess the state of the art; they reflect a description of scientific progress, an inventory of gaps in the current set of testing guidelines, recommendations of guidelines that need updating, and proposals for developing or updating guidelines.

The specific testing requirements developed by EPA and OECD are assumed to have a sound scientific foundation and are generally accepted by interested stakeholders. As indicated, this chapter discusses the consensus protocols focusing primarily on EPA guidelines. It has been organized to present the more general toxicity tests first and then the tests designed to evaluate specific toxicity end points. Thus, the toxicity tests characterized by exposure duration—acute, subchronic, and chronic—are reviewed first; these tests are designed to gain an understanding of systemic effects, given various lengths of exposure, and can be used to guide human health risk assessment for those exposure durations. Toxicity tests designed to evaluate specific end points are discussed next and include tests for reproductive and developmental toxicity, neurotoxicity, immunotoxicity, and genotoxicity. It is important to note that some specialized end points are evaluated by various clinical measures or histopathology conducted in the exposure-duration tests. Results of general toxicity tests often indicate a need to conduct more specialized tests. The chapter concludes with a discussion of metabolism and pharmacokinetic studies. The intent of this chapter is to provide an overview of the rationale for conducting specific toxicity tests, the basic aspects of the study protocols, and the possible shortcomings of currently accepted tests. The descriptions are meant not to be exhaustive but simply to provide a context for evaluating toxicity-testing strategies. Detailed descriptions of study protocols can be found in the cited references.

---

[1]See http://www.oecd.org/document/30/0,2340,en_2649_34377_1916638_1_1_1_1,00.html for a listing of the OECD monographs.

TOXICITY TESTING CHARACTERIZED
BY EXPOSURE DURATION

Acute Toxicity Testing

Acute toxicity tests evaluate the adverse effects of short-term exposure and are considered by EPA to be an "integral step in the assessment of [a chemical's] toxic potential under the regulatory framework of its pesticide and toxic substances programs" (EPA 1998a). To be considered an acute exposure, dosing may be done once or may be done several times within or continuously throughout a 24-hour period, but use of a single dose is by far the most common method. The test animals, typically rodents (rats or mice) are observed for a period of several days to 2 weeks after dosing, and observations of deviant behavior, growth, or mortality are recorded. Historically, the primary focus of an acute toxicity test was to determine a chemical's median lethal dose (LD_{50}), the dose that causes death in 50% of the test animals. Today, acute toxicity tests are used also to determine dosing regimens for longer-term toxicity tests and to evaluate more fully the effects of acute exposure.

Acute testing protocols have evolved over the years to conserve animal use, to minimize the pain and discomfort of the test animals, and to obtain more information on the pathogenesis of toxicity. If a chemical is judged to have low toxicity, a limit test is first conducted. The limit test is a sequential test that uses a maximum of five animals with a starting test dose of 5,000 mg/kg (EPA 1998b). If three or more animals survive, the LD_{50} is considered to be greater than 5,000 mg/kg, and no further testing is conducted. If the substance proves to be more toxic than expected (that is, three or more animals die), the primary test recommended by EPA to assess acute oral toxicity is the up-down procedure (UDP) (EPA 1998b). The UDP uses one animal per exposure, and the animals are dosed sequentially at 48-hour intervals. The first animal is dosed a step below the best estimate of the LD_{50}. If the animal survives, the second animal receives a dose that is higher by a factor of 3.2; if the first animal dies or appears moribund, the second animal receives a dose that is lower by a factor of 3.2. This process continues until death is observed or an upper bound is reached (usually 2,000 or 5,000 mg/kg). EPA has developed a software program that incorporates the data obtained from the UDP to calculate the LD_{50} and the confidence interval.

Although the preceding discussion focused on oral exposure, the route most relevant to potential human exposure (oral, inhalation, or ermal) is typically evaluated. Acute testing protocols are available for inhalation and dermal exposure (EPA 1998c,d). EPA has developed toxicity categories on the basis of LD_{50} or median lethal concentration (LC_{50}) values (see Table 2-1). OECD (2001a) has a similar ranking system. EPA uses the categories to determine precautionary labeling requirements, personal protective equipment requirements, and restrictions on entry into pesticide-treated areas.

Acute toxicity data have benefits beyond toxicity ranking. Acute studies reveal whether frank toxicity is sudden, delayed, time-limited, or continuous. The time to onset and resolution of toxicity can provide insight into the time course of absorption, distribution, and clearance of a toxicant. Acute toxicity data can provide some idea of relative bioavailability by comparing data on various routes of exposure and can provide information on clinical signs potentially relevant for physicians who are treating patients and for scientists who are developing hypotheses about pathogenesis and target organs affected by acute exposures. That is especially important because toxic effects of acute exposure are often different from those of prolonged lower-level exposure. As discussed in greater detail in Chapter 6, acute toxicity tests can be redesigned to provide additional information on more subtle effects than lethality and gross clinical signs. One particular end point that has received increasing attention is cardiovascular toxicity, specifically adverse effects on ion channels in the myocardium that lead to abnormalities in the electrocardiogram, namely prolongation of the QT interval. Changes in the QT interval have been linked with cardiac arrhythmia that can progress to more serious cardiac events, including failure. However, the link has not yet been proven, and many believe that more research is needed on an alternative indicator of cardiac arrhythmia. The pharmaceutical industry

TABLE 2-1 EPA Acute-Toxicity Categories

Study	Category I	Category II	Category III	Category IV
Oral LD_{50}	≤50 mg/kg	>50-500 mg/kg	>500-5,000 mg/kg	>5,000 mg/kg
Dermal LD_{50}	≤200 mg/kg	>200-2,000 mg/kg	>2,000-5,000 mg/kg	>5,000 mg/kg
Inhalation (4-h) LC_{50}	≤0.05 mg/L	>0.05-0.5 mg/L	>0.5-2 mg/L	>2 mg/L

Source: EPA 1998a.

does evaluate effects on cardiovascular function of potential drug candidates as part of its regulatory process; EPA does not have formal guidelines for evaluating cardiovascular toxicity.

Acute toxicity studies are based on the assumption that acute toxicity and lethality in animal models are relevant to humans. For many chemicals, the experience in humans is inadequate to confirm that assumption, but enough examples support it to continue this mode of hazard assessment. However, dose extrapolations from animals to humans are not simple: smaller rodents generally have a far greater rate of metabolism than do humans and therefore clear a chemical more rapidly, reducing total exposure relative to dose. Extrapolations therefore use plasma or tissue concentrations, an adjustment or uncertainty factor, or, as a surrogate for metabolic rate, doses calculated on the basis of body surface area or a quantity equal to body weight raised to the ¾ power. Moreover, metabolic and biologic differences sometimes lead to responses in animals or humans that are absent in the other species, termed species specificity. Knowledge of species differences in toxic responses is critical for extrapolating from animal data to human risk.

The scientific consensus remains that assessment of acute toxicity can help scientists to evaluate and manage the risks associated with potential exposure to noxious agents. Acute toxicity tests provide at least one relatively quick and inexpensive tool in testing schemes that screen large numbers of chemicals and identify chemicals that warrant further toxicity testing.

Subchronic or Repeated-Dose Toxicity Testing

Subchronic studies evaluate the adverse effects of continuous or repeated exposure over a portion of the average life span of experimental animals. They provide information on target-organ toxicity and bioaccumulation potential and are designed to determine no-observed-adverse-effect levels (NOAELs), which are used to establish standards or guidelines for human exposure. Subchronic studies are not designed to assess effects that have a long latency period, such as cancer, but do provide information that can be used in setting doses for chronic toxicity and carcinogenicity studies.

The exposure durations for subchronic studies are typically 28 or 90 days (see Appendix B for a list of EPA and OECD guidelines). Administration of the chemical (oral, inhalation, or dermal) is usually determined by the route of potential or actual human exposure. Depending on exposure duration, animals are often observed for 2 or 4 weeks after the end of treatment for reversibility, persistence, or delayed occurrence of adverse effects.

In 90-day studies, 20 rodents (10 of each sex) or eight nonrodents (four of each sex) are used for each dose group and the control group. Additional animals are included in the control and high-dose groups if satellite groups are used to evaluate effects after termination of treatment. In some cases, the shorter-term studies are conducted with fewer animals, such as five rats per sex per dose, and may evaluate fewer measures than the 90-day studies.

Typically, doses in subchronic studies are selected to define a dose-response relationship. The lowest dose should produce no adverse effects, the highest dose should induce toxic effects without compromising survival or inducing severe suffering, and the intermediate dose should produce a gradation of effects. A control group is also included. When a dose of 1,000 mg/kg per day in oral or dermal studies or 1 mg/L in inhalation studies is not toxic, further dosing above these quantities is not required. Oral dosing occurs daily if test material is incorporated in food or water or 5 days/week if the test material is administered by gavage (the method typically used for rodents) or capsule (typically used for dogs). In inhalation studies, exposure is usually conducted for a period of 6 hours/day for 5 or 7 days/week. Test guidelines require measurement and evaluation of a number of parameters, including clinical signs (such as changes in skin, fur, eyes, secretions, gait, posture, and response to handling), motor activity, grip strength, sensory reactivity to stimuli, body weight, food consumption, clinical pathology (clinical chemistry and hematology), and ophthalmology. At study termination, a gross necropsy is conducted on all animals, and selected organs are weighed. A full histopathologic analysis is conducted on all animals in the control and high-dose groups, on all animals that were killed or died during the study, and on all gross lesions. Target organs are examined in all animals. Statistical methods are used to evaluate the data.

Subchronic studies can provide initial or definitive data for risk-assessment purposes. However, the studies are sometimes limited by the smaller sample size, which reduces the sensitivity of the study to detect adverse effects. They often provide the basis of dose selections for longer-term studies, including chronic toxicity and carcinogenicity studies.

Chronic Toxicity and Carcinogenicity

The purpose of chronic toxicity testing is to determine the cumulative adverse effects of repeated daily oral, dermal, or inhalation exposures of test animals to various doses of a chemical for at least 12 months (EPA 1998e). The purpose of carcinogenicity testing is to determine the cumulative neoplastic effects of repeated daily oral, dermal, or inhalation exposures to various doses of test chemicals over most of the life span of the test species (EPA 1998f). EPA provides separate guidelines for chronic toxicity and carcinogenicity, but testing is most often combined for these two end points (EPA 1998g).

EPA guidelines (EPA 1998e) for chronic toxicity specify that "testing should be performed with two mammalian species, one a rodent and the other a nonrodent. The rat is the preferred rodent species and the dog is the preferred nonrodent species." Other species can be used with adequate justification. Dose selection is generally based on results of a 90-day study; the highest dose should be the one that causes only mild signs of toxicity and does not alter the length of the study. The intermediate dose is chosen to produce a gradation of toxic effects, and the lowest dose should produce no evidence of adverse effect and thus should allow determination of a NOAEL. At least three dose groups and a control group should be included with 40 rats (20 of each sex) or eight dogs (four of each sex) in each group. EPA guidelines state that body weights and food consumption should be measured and that clinical pathology (hematology, clinical chemistry, and urinalysis) should be conducted at specified intervals during the study. At the end of the study, all animals should be subjected to gross necropsy, weights of major organs should be determined, and all gross lesions and tissues and organs of the digestive system, nervous system, glandular system, respiratory system, cardiovascular and hematopoietic system, and urogenital system should be preserved for histopathologic examination. Ophthalmologic examinations are also recommended. A full histopathologic analysis should be conducted on all controls and animals in the high-dose group and on gross lesions. If exposure-related changes are detected, the analysis is extended to all treatment groups (EPA 1998e).

Carcinogenicity bioassays are conducted with rodents, typically rats and mice, for a minimum of 24 months (rats) and 18 months (mice) and are designed to provide data for cancer-hazard identification and dose-response evaluation. Dose-selection guidelines are similar to those for the chronic toxicity studies; however, group sizes are larger (50 rodents of each sex per group), and clinical pathology involves examination of blood smears. At the end of the study, gross necropsy and histopathology are extensive because the primary focus is on detecting neoplasms. The National Toxicology Program (NTP) has conducted over 600 lifetime cancer bioassays and has been at the forefront of developing definitive guidelines for detecting carcinogenic activity in rodents; the carcinogenicity data obtained reside in a public database.

EPA guidelines for combined chronic toxicity and carcinogenicity testing (EPA 1998g) combine testing for chronic toxicity and carcinogenicity summarized above. In a combined test, the two species typically used are rats and mice—rats mainly for dosing by oral and inhalation routes and mice for the dermal route. Three dose groups and a control group are used, and at least 100 animals (50 of each sex) are used for each group. Additional animals—at least 20 (10 of each sex)—are included at each dose and in the control group as satellite groups for determination of chronic toxicity after 12 months; end points similar to those described for the chronic toxicity test are used. The minimal duration of daily exposure is 2 years for rats and 18 months for mice, and end points similar to those described for the carcinogenicity test are examined at the end of the study.

Considerable effort is being devoted to developing alternative transgenic and knockout animal models for carcinogenicity testing in Europe and the United States. The goal is to develop models that will increase the sensitivity of detection of carcinogenic lesions and shorten the time for their appearance; the latter would have the effect of conserving the resources required to test each agent and increase the number of agents that can be tested. Some of the efforts are being coordinated through the International Conference on Harmonization (ICH) Expert Working Group on Safety. That group, in collaboration with the Health and Environmental Sciences Institute (HESI) of the International Life Sciences Institute (ILSI), conducted an evaluation of six animal models for their ability to detect the effects of a group of 21 chemicals, which included genotoxins and carcinogens. The results of those efforts were discussed at a workshop (Cohen et al. 2001) and presented in a special issue of Toxicologic Pathology (Vol. 29, supplement issue, 2001), which also presented detailed information on the models. The conclusion drawn from the evaluations was that some of the models might have use in hazard identification, providing information similar to that obtained from the 2-year combined chronic toxicity and carcinogenicity bioassay, in conjunction with data from other sources in a weight-of-evidence approach to risk assessment. Determination of the usefulness of the models is still limited by the amount of comparative data available. Considerable effort is being devoted to broadening the comparison of tumor data from transgenic mouse strains and strains of mice traditionally used in lifetime bioassays.

TOXICITY TESTING CHARACTERIZED BY SPECIFIC END POINT

Toxicity testing of most chemicals begins with acute testing, progresses to subchronic testing, and, depending on the results, concludes with chronic testing. Evaluation in those studies may indicate the need to obtain more information on specific toxicity end points. The following sections discuss the tests used to evaluate reproductive and developmental toxicity, neurotoxicity, immunotoxicity, and genotoxicity. In vitro tests for cytotoxicity and other end points are also briefly discussed.

Reproductive and Developmental Toxicity

Reproductive and developmental toxicity testing includes a broader category of end points than other kinds of toxicity testing because of the multiple stages of exposure and the variability of possible effects. Exposures of sexually mature animals can result in sterility or decreased fertility by depleting or affecting ova or sperm or by affecting endocrine functions of organs involved in reproduction. If fertilization occurs, abnormalities of ova and sperm can result in embryonic death, failure of implantation, congenital malformations, embryonic growth retardation, genetic disease, or cancer in the offspring. Exposures during pregnancy can result in embryonic or fetal death, congenital malformations, reversible or irreversible growth retardation, or premature or delayed parturition; they may also have delayed postnatal effects, such as cancer, neurobehavioral effects, growth retardation, and death. Toxicant exposures of neonatal, immature, or adolescent organisms may result in growth retardation or stimulation, endocrine abnormalities, immunologic deficits, neurobehavioral effects, cancer, or death.

The general purpose of reproductive and developmental toxicity assays is to evaluate the competence of breeding pairs to produce phenotypically normal offspring. All or most of the reproductive cycle is evaluated. Four types of reproductive and developmental studies are discussed here—screening-level reproductive-toxicity assays, prenatal developmental-toxicity and teratology studies, generational tests, and reproductive assessment with continuous breeding. These assays are conducted because of their assumed relevance for predicting human hazard potential, but the data from such models may or may not be relevant for predicting human risk. Thus, the predictive power of the tests may be limited by differences in the underlying biology. A famous example of how species differences can be important is developmental exposure to thalidomide, to which rats are highly resistant and humans are exquisitely sensitive.

The assays described are apical tests—complex experiments that measure complicated end points, each of which is an integrated measure of multiple facets of the machinery necessary for successful reproduction and development. Apical tests provide little insight into the hundreds of molecular events, mechanisms, and targets responsible for toxicant action. Although they are useful for determining whether there is an overall effect, the lack of mechanistic insight is an important limitation. Future advances in testing will probably rely on our ability to discern the individual biologic underpinnings of toxicity, a complicated task in this setting.

Screening-Level Reproductive-Toxicity Assays

In these assays, animals are dosed with the test chemical for at least 2 weeks before mating and then for a maximum of 2 weeks of breeding. The females are dosed through gestation, and the test is terminated on postnatal day 4. The measurements made provide insight into gonadal function, fertility, pregnancy, parturition, and prenatal and postnatal developmental toxicity. OECD testing guidelines (TG) 421 and 422 (OECD 1995a, 1996) are reproductive and developmental screening tests; however, TG 422 (OECD 1996) is a combined repeated-dose toxicity study in combination with the reproductive and developmental screening test. These are screening-level assays used to make decisions about the need for further testing as part of the OECD screening information dataset (SIDS) program.

Prenatal Developmental-Toxicity and Teratology Studies

The prenatal developmental-toxicity study (OECD 2001b) is used to examine embryonic and fetal toxicity as a consequence of exposure during pregnancy (for example, growth retardation, anatomic variations, teratogenicity, and lethality). Young mature virgin females are artificially inseminated or mated. The time of mating is noted, and groups of pregnant animals are either untreated or treated with three different doses of the test agent. In OECD TG 414 (OECD 2001b), the treatment is given from the time of implantation to scheduled cesarean section. If preliminary studies do not indicate a high potential for preimplantation loss, treatment may be extended to include the period from mating to the day before the scheduled cesarean section. The day before expected birth, the uterus is removed by cesarean section, and the uterus and fetuses are examined. If dosing is initiated before or at the time of implantation, preimplantation loss is evaluated.

Generational Tests

The prototypical reproductive-toxicity assay is the one-generation test (OECD 1983), although this test is not included by EPA in its test guidelines. The test chemical is administered to young adult rats of both sexes (generally to breeding pairs) during a prebreeding period covering one spermatogenic cycle and the last two stages of oocyte maturation and during mating. Dosing of females continues through pregnancy and nursing (3 weeks after birth). Pups are evaluated from birth through weaning, and birthweight, postnatal growth, survival, litter sizes, and sex distribution are recorded. Adult males typically are killed after the mating period, and sperm production and quality are assessed. Reproductive organs of both sexes of the parental generation are assessed grossly and histologically. The assay includes assessment of gonadal function, estrous cycling, mating behavior, fertility, parturition, and lactation in the parental animals and prenatal and postnatal development in the offspring. The assay has been modified over time, most recently to include end points that are sensitive to endocrine-disrupting chemicals.

The results of the one-generation reproductive-toxicity test in rodents are often used in risk assessment. In chemical regulation in Europe, it is part of a tiered testing system in which a two-generation test may follow the one-generation test. The two-generation test is most often triggered when a specified production volume is reached. The ICH has published a guideline on reproductive-toxicity testing of medicinal products. The guideline describes a flexible design in which a reproduction and fertility test comparable with the one-generation reproductive-toxicity test can be run as a stand-alone assay or as part of the developmental-toxicity and perinatal and postnatal assessment of new drugs. Although there are differences from the OECD one-generation protocol, the general principle of the test is comparable.

The design of the two-generation test (OECD 2001c) is in principle similar to that of the one-generation test, but the first generation of offspring (F1) is followed through sexual maturation and the production of a second generation. The treatment of the parental generation is equivalent to that in the one-generation test. Data on sperm quality and estrous cycling in the offspring are collected, and the offspring are observed for developmental milestones, including some behavioral measures and histopathologic characteristics of sex organs, brain, and other potential target organs. Direct dosing of the F1 animals begins at weaning and is continuous through the end of the test. On reaching sexual maturity, these animals are bred. The F2 generation is evaluated through weaning. The data collected are similar to those in the one-generation study. The two-generation test is considered the appropriate test for reproductive toxicity, and the OECD test guideline has recently been updated to reflect the scientific state of the art (OECD 2005). EPA lists only the two-generation test as its reproductive toxicity assay (EPA 1998h).

Reproductive Assessment by Continuous Breeding

The protocol for reproductive assessment by continuous breeding has been conducted almost exclusively by the NTP. Breeding pairs cohabit for an extended period (14 weeks), during which they are continuously exposed to the test agent. Each litter produced is examined and then discarded. This study design allows the determination of the maximal number of litters that can be produced. If effects on fertility are noted, additional study legs can be run to determine which sex is affected and to generate hypotheses regarding the mechanism of toxicity. The test can also be extended into a two-generation-like protocol. The assumptions and uncertainties are similar to those of the one-generation test. This study design has not been required as part of the regulatory assessment process for pesticides, but the data are useful.

Neurotoxicity

Neurotoxic effects in animals and humans can be assessed with a wide array of methods, including neurochemical, anatomic, physiologic, and behavioral. For example, neurochemical effects of an agent can include selective effects on synthesis, reuptake, release, or metabolism of specific neurotransmitters. Anatomic changes can include alterations of the cell body, the axon, or the myelin sheath of neurons or of the thickness of cell layers in specific brain regions. At the physiologic level, a chemical might reduce the speed of neurotransmission or change the thresholds for neural activation. Behavioral alterations can include changes in sensations of sight, hearing, or touch; alterations in simple or complex reflexes and motor functions; alterations in cognitive functions, such as learning, memory, and attention; and changes in a wide array of psychologic and social behaviors (WHO 2001).

Regulatory agencies consider data from required animal toxicology studies and effects reported in the published literature when evaluating the neurotoxic potential of chemicals. This section briefly summarizes the types of neurotoxicity evaluations that are conducted in animals and recommended in toxicity-test guidelines by EPA and OECD (see Appendix B for a summary of the guidelines). More comprehensive reviews of the available neurotoxicity tests have been published by the National Research Council (NRC 1992), EPA (1998i), the World Health Organization (WHO 2001), and OECD (2004c).

Neurotoxicity Testing in Standard Toxicity Studies

Neurotoxicity evaluations required by EPA and OECD guidelines for standard acute, subchronic, and chronic toxicity tests include detailed clinical observations, functional tests, and histopathology. Detailed clinical observations are made outside the home cage, preferably in a standard arena, and at similar times on each occasion. Observations should include evaluation of skin and fur, eyes, and mucous membranes; respiratory and circulatory effects; autonomic effects, such as salivation;

central nervous system effects, including tremors and convulsions; level of activity; gait and posture; reactivity to handling or sensory stimuli; altered strength; stereotypies; and bizarre behavior, such as self-mutilation and walking backward. Observations should be detailed and carefully recorded, preferably with scoring systems explicitly defined by the laboratory. Toward the end of the repeated-dose standard adult toxicity studies, functional tests are required, including an assessment of motor activity, grip strength, and sensory reactivity to stimuli of different types, such as visual, auditory, and proprioceptive stimuli. Brain weight and histopathologic characteristics are required. Specifically, multiple sections of the brain should be examined (including cerebrum, cerebellum, medulla, pons, and pituitary), as should sciatic and tibial nerves close to muscle, specimens of three levels of the spinal cord (cervical, midthoracic, and lumbar), and eyes, including the retina and optic nerve. The clinical and functional end points mentioned here can be altered by specific and nonspecific effects on the nervous system, especially at maximum tolerated doses that can cause substantial systemic toxicity.

The standard developmental study (EPA 1998j) includes gross pathologic assessment of the nervous system. Two-generation reproduction studies (EPA 1998h) are used to evaluate clinical signs of toxicity and brain weight in offspring. Those evaluations can provide an initial indication of potential neurotoxic effects after postnatal exposures but are much more limited than those required in the standard adult toxicity studies.

Adult Neurotoxicity Studies

OECD TG 424 (OECD 1997) and the EPA neurotoxicity screening battery (EPA 1998k) are similar. Both include detailed clinical observations or a functional observational battery in the home cage and open field; functional tests, including assessments of motor activity, grip strength, and reactivity to sensory stimuli; and neuropathologic examination of perfusion-fixed tissues. Adult neurotoxicity studies require functional tests and clinical observations similar to those in standard toxicity studies but require perfusion-fixed tissues, more frequent measurement of functional tests, and observations to be conducted without knowledge of treatment level.

In general, OECD and EPA neurotoxicity tests may be required when there are structure-activity concerns or when neurotoxic effects have been revealed in standard toxicity, reproductive, or developmental studies. The OECD neurotoxicity TG recognizes the redundancy of standard toxicity studies, and TG 424 encourages flexibility to minimize the number of end points that merely duplicate those of standard repeated-dose toxicity studies.

Developmental Neurotoxicity Studies

The developmental-neurotoxicity study protocol (EPA 1998l) is designed to develop data on the potential functional and morphologic hazards to the nervous system in offspring of mothers exposed during pregnancy and lactation. OECD (2004b) has developed similar draft guidelines for a developmental-neurotoxicity test. The EPA guidelines require that pregnant females be dosed from gestational day (GD) 6 through postnatal day (PND) 10. Recently, EPA extended the dosing period from GD 6 through PND 21 (that is, until weaning). Motor activity is measured repeatedly on PND 13, 17, 21, and 60. Auditory-startle habituation is measured around weaning and on PND 60. Auditory-startle habituation, as conducted in this study, is primarily a measure of reactivity to repeated loud bursts of noise (for example, 120 dB for 10 msec). A test of learning and memory is also required around weaning and on PND 60. The EPA developmental-neurotoxicity test guidelines (EPA 1998l) require periodic clinical observations of the dams and pups with standardized procedures by trained technicians who are unaware of the animals' treatment. Interobserver reliability is required if more than one observer is used in a given study. The EPA guidelines (EPA 1998l) require detailed neuropathologic evaluation, including measurements of immersion-fixed brain taken on PND 11. EPA allows neuropathologic evaluation at PND 22 because the exposure period was extended through PND 22, but the pups should be perfusion-fixed. At the termination of the study (usually on PND 60), the pups are perfusion-fixed, and the central and peripheral nervous systems are evaluated according to the EPA neurotoxicity-screening battery guideline with the additional requirement of measurements of the brain.

An important limitation in using the rat model for developmental-neurotoxicity testing is that there are important species differences in brain development relative to birth. In general, brain development in rats from birth to about PND 11 is roughly equivalent to brain development in human fetuses during the third trimester of gestation (Rice and Barone

2000). Thus, it could be difficult to experimentally reproduce potential environmental exposure to the human fetus in rat developmental-neurotoxicity studies. For example, exposure through maternal rat milk may be substantially different from human in utero exposures in terms of relative amounts of parent compound and metabolite. But direct dosing of pups during early lactation may not necessarily reflect human in utero exposure after dietary, dermal, or inhalation exposure to pregnant females. Measurement of parent compound and toxic metabolite in milk and evalution of biomarkers of exposure or effect after exposure can potentially be used in physiologically based pharmacokinetic and pharmacodynamic models to characterize dose-response relationships and improve extrapolation of results from animal studies to humans (Dorman et al. 2001). Although studies of lactational or placental transfer could prove valuable for interpreting toxicity-study results, such study designs are less well established under conditions in which human data will not be available, the mode of action is not well characterized, and the toxic component has not been identified. Therefore, an ILSI-HESI committee evaluating use of pharmacokinetic and metabolism data for developmental and developmental-neurotoxicity testing recently concluded that more experience is needed on how best to carry out these studies before pharmacokinetic data on fetus, offspring, and maternal milk are routinely required (Barton et al. 2005).

The developmental-neurotoxicity test is one of the most logistically difficult EPA-guideline tests to conduct and requires specialized expertise in neurobehavioral testing and morphometric analysis. Historical control data from laboratories conducting EPA developmental-neurotoxicity guideline studies can be variable, particularly at the earlier times of PND 13 and 17 (Raffaele et al. 2003, 2004; Sette et al. 2004). Some of the variability could be reduced by improving environmental experimental conditions and methods, as discussed by Garman et al. (2001) and Cory-Slechta et al. (2001). However, the variability may also be due, in part, to normal variability during a period of rapid development, to the practical definition of birth date that could span 23 hours and does not take into account gestational age, and to the practical necessity of testing large numbers of animals over several hours during the day and across multiple days of testing depending on when the pups were born (Li 2005).

Initial analysis of the relative sensitivity of developmental-neurotoxicity testing compared with other standard end points and studies indicates that, in general, the developmental-neurotoxicity study is not more sensitive than chronic bioassays and other reproductive-developmental end points but can provide additional characterization of potential neurotoxic effects after developmental exposures (Makris et al. 1998; EPA 1999; Middaugh et al. 2003).

Specialized Studies for Neurotoxicity Testing

EPA has developed six test guidelines for neurotoxicity testing (see Appendix B). The neurotoxicity-screening battery (EPA 1998k) and developmental-neurotoxicity study (EPA 1998l) were discussed above. The delayed-neurotoxicity test is typically required for organophosphorus substances and includes behavioral, histopathologic, and neurochemical assessments in the hen (EPA 1998m). The remaining guidelines for schedule-controlled operant behavior (EPA 1998n), peripheral-nerve function (EPA 1998o), and sensory-evoked potential (EPA 1998p) are outlined briefly in this section. A more comprehensive review of neurotoxicity end points can be found in EPA's neurotoxicity risk-assessment guidelines (EPA 1998i).

The test guideline for schedule-controlled operant behavior requires that subjects be trained until they display demonstrable stability in performance before exposure. This guideline is designed to evaluate performance of a learned behavior and not learning or memory itself. EPA states that substances that have been observed to produce neurotoxic signs in other toxicity studies (such as central nervous system depression or stimulation) and substances that are structurally similar to neurotoxicants that affect performance, learning, or memory may be appropriate to evaluate with this test. Although schedule-controlled operant behavior testing may be useful to more fully characterize the potential neurotoxicity of a chemical, it has not been found to be more sensitive than a functional observational battery or an assessment of motor activity in several independent laboratories for different classes of chemicals (Moser et al. 2000).

The peripheral-nerve test function is used to evaluate peripheral-nerve conduction velocity and amplitude in anesthetized animals with electrophysiologic techniques. EPA indicates that substances that have been shown to produce related effects in other studies (such as neuropathologic changes in peripheral nerves) and substances with a structural similarity to those causing peripheral neuropathy may be appropriate to evaluate with this test.

The test for sensory evoked potentials is also an electrophysiologic test that evaluates the effects of chemicals on brain electric potentials after stimulation of the visual, auditory, or somatosensory system. The test is recommended if there is reason to believe that particular sensory functions are specifically sensitive to the test compound.

In contrast with EPA, OECD did not develop specific test guidelines for more specialized neurotoxicity evaluations. Instead, OECD developed a guidance document for neurotoxicity testing that provides general descriptions, references, and commentary for a wide variety of behavioral, neurologic, neurochemical, neurophysiologic, and morphologic techniques (OECD 2004c). The OECD neurotoxicity guidance document emphasizes an "iterative" testing strategy that includes an evaluation of the degree of concern about the neurotoxic effects and the possible concentrations to which people may be exposed to determine the adequacy of the existing data to evaluate potential risk.

OECD indicates that initial animal data for neurotoxicity assessment are most often provided by standard single-dose (OECD 1981, 1987, 2001d,e,f) or repeated-dose toxicity studies (OECD 1995b, 1998) in which functional histopathologic information is gathered on all major organ systems, including the nervous system. If information indicates possible neurotoxic effects, OECD recommends that additional end points be included in the initial standard tests to obtain in-depth information about specific neurotoxic effects. However, OECD cautions that the decision to add specific end points to the initial study should take into consideration the potential for confounding toxicologic effects of the higher doses usually required in the initial studies. For example, excessive systemic toxicity at doses near the maximum tolerated dose can cause indirect effects on the nervous system that confound interpretation of tests of learning and memory.

In summary, a tiered or iterative approach to neurotoxicity testing is being used by OECD and EPA. Initial evaluation of the nervous system can be obtained with standard toxicity studies, which are typically conducted over a range of doses that include the maximum tolerated dose. OECD encourages an iterative approach that includes a determination of whether the neurotoxic effects and exposure assessment provide sufficient data for an assessment of risk (OECD 2004c). The available information on the test chemical and on structurally related chemicals can be used to guide selection of additional neurotoxicity tests.

Immunotoxicity

The immune system is responsible for defending the body against infection by viruses, bacteria, and other disease-producing microorganisms. It also plays a role in cancer surveillance, destroying cells that have become transformed and might otherwise develop into tumors. For the immune system to recognize the wide array of pathogens that are present in the environment, it relies on many cell types that play mutually supporting roles in generating an immune response. Those cells arise from stem cells in the bone marrow and thymus and are found throughout the body in lymphoid tissues and in the blood as white blood cells. Because of the complexity of the immune system, assessing the influence of chemical exposure is a complicated and difficult task.

Two main types of immunotoxic responses may result from xenobiotic exposure: immunosuppression (in which one or several specific functions of the immune system are inhibited, and the inhibition results in increased susceptibility to infections or tumors) and antigenicity (in which the immune system recognizes the xenobiotic as foreign and mounts an immune response to it). Antigenicity is more commonly known as allergy, which is often referred to as hypersensitivity even though it constitutes a normal immune response to a foreign substance. Contact hypersensitivity is a relatively common cutaneous immunotoxic response that depends on the sensitization of T lymphocytes and leads to the development of inflammatory lesions in the skin on re-exposure to the sensitizing chemical. Alternatively, IgE-antibody-mediated allergic responses occur immediately on re-exposure to a sensitizing chemical, and symptoms depend on the route of exposure to the xenobiotic. If the exposure occurs by inhalation, asthmatic conditions, such as wheezing or pulmonary congestion, may arise; if the exposure is oral, there may be an intestinal reaction of diarrhea or vomiting; and systemic effects may include decrease in blood pressure, vessel leakage, or shock.

A third, less studied immunotoxic effect of xenobiotic exposure is autoimmunity. In this situation, the xenobiotic may act as a partial antigen (a hapten) and create a new antigen by binding to tissue protein. If the immune system recognizes the new tissue-associated protein as foreign, an immune response to the tissue is generated and results in an autoimmune response and, if it is persistent, autoimmune disease.

Potential immunosuppressive effects are often identified during subchronic toxicity testing. Whole and differential blood-cell counts, lymphoid organ weights, and histopathologic examination of the spleen, thymus, lymph nodes, and bone marrow can indicate effects on the immune system. However, those end points alone may not always be predictive of immunotoxicity (Luster et al. 1992, 1993). Because test animals are routinely maintained under very clean housing conditions, the immune system can essentially be considered to be "resting" unless specifically challenged with antigen. Therefore, a specific immune response must be induced to assess the ability of a xenobiotic to cause immunosuppression.

The functional immune test recommended by EPA is the in vivo antibody response to sheep red blood cells (SRBCs) (EPA 1998q). The test is done in mice and rats unless the pharmacokinetic data are similar in both species, in which case either species may be used. The animals are exposed to test and control compounds for at least 28 days and are then immunized with SRBCs. At the end of the exposure period (4-5 days after SRBC injection), either a plaque-forming cell assay to measure the number of B cells making anti-SRBC antibody (immunoglobin M, IgM) or an enzyme-linked immunosorbent assay (ELISA) to measure serum anti-SRBC IgM is performed. Additional or followup tests may include flow cytometry to evaluate subpopulations of T and B cells if there is substantial suppression of the anti-SRBC response or a functional test for natural killer cells to assess the chemical's effect on nonspecific immunity (EPA 1998q). The Food and Drug Administration (FDA) also recommends the SRBC assay as a test for immunosuppression if pharmacokinetic studies indicate that an investigational new drug or its metabolites concentrate in immune tissues (FDA 2002).

Immunotoxicity testing for contact hypersensitivity has been conducted for many years; guinea pigs have been the model of choice. The guinea pig maximization test (GPMT) and the Buehler test (BT) have been validated for use in screening for skin sensitizers. The BT and GPMT involve delivery of a test substance or vehicle to the skin of guinea pigs either topically (BT) or by intradermal injection with and without Freund's adjuvant (GPMT). In both tests, skin inflammatory reactions are graded and recorded after challenge doses of the test substance.

The expense and technical difficulties associated with guinea pig tests led to a concerted effort among interested parties to develop and validate a mouse model that could replace the guinea pig tests. The mouse local lymph node assay (LLNA) has now obtained regulatory acceptance for screening for skin sensitizers The LLNA detects proliferation of lymphocytes in lymph nodes on application of potential skin sensitizers to the ears of mice. The test affords a quantitative measure of cell proliferation as a function of incorporation of radioisotope into DNA of dividing lymphocytes. The LLNA is considered a refinement of traditional guinea pig skin sensitization assays because it minimizes pain and distress of animals and uses fewer animals (EPA 2003; FDA 2002; NTP 1999). Alternatively, the mouse ear swelling test may be used to detect moderate to strong sensitizers; if an agent is positive, it may be designated a potential sensitizer without further testing in guinea pigs (EPA 2003).

Tests for immunotoxicity are also outlined in guidelines for microbial-pesticide toxicity studies (EPA 1996). For those agents, EPA considers reporting of observed allergic responses of humans to microbial pest-control agents (MPCAs) sufficient to address potential health concerns about allergy or hypersensitivity. If there is a potential for a virus-containing MPCA to cause immunodeficiency in mammals, specific followup tests are prescribed case by case.

Genotoxicity

Genotoxicity refers to adverse effects on DNA, genes, and chromosomes, and genetic toxicology is the branch of toxicology that studies those effects. There is concern about such effects because many human diseases are of mutational origin. Down, Klinefelter, and Turner syndromes are the most frequently encountered chromosomal aberrations in humans, and retinoblastoma (a gene mutation) occurs in one of 20,000 births (Flamm et al. 1977). Brusick (2001) provides information on the effects of mutagens on the human gene pool.

The purpose of conducting genotoxicity studies is to identify agents that have the potential to alter DNA. Originally, the focus of genetic toxicology was on whether the effects could be transmitted to future generations. To evaluate that possibility, methods were developed to determine the transmissibility of genetic damage to progeny. Those methods used whole animals, primarily rats or mice. Over the years, the concern about what the effects mean for potential carcinogenicity has increased. One could argue that most studies as now practiced are directed toward evaluating carcinogenic potential (FDA 2000a). As a result, more in vitro testing for genotoxicity is done. Animal studies, however, remain a part of test guidelines of EPA, FDA, and other government agencies and organizations.

Test methods in genetic toxicology are categorized according to their ability to detect gene or point mutations; chromosomal effects, such as breaks, gaps, translocations, and aneuploidy (loss or gain of one or more chromosomes); or other DNA damage, as can be indicated by un-scheduled DNA synthesis (Herbold et al. 2001). There are over 50 specific tests of genetic toxicity; they are conducted in bacteria, yeast, fruit flies, mammalian cells, and whole animals. Table 2-2 lists representative genotoxicity tests, which include both in vitro and whole-animal tests, particularly in the case of chromosomal abnormalities. Only the tests that are used most often for regulatory purposes are discussed here.

With the exception of the heritable-translocation assay and the dominant-lethal assay, the tests discussed here have a common limitation. They provide information only on carcinogenic or mutagenic potential (that is, hazard identification) and are less useful for providing quantitative information needed for risk assessment. That glaring deficiency is most apparent when one considers how to assess the risks to humans from chemicals that cause gene mutations. The only assay that can be used is the specific locus assay conducted in mice. However, that assay is not typically used, because it is costly and requires the use of large numbers of animals.

Gene or Point Mutations

A variety of methods that use bacteria or mammalian cells have been developed. Because of costs, few studies are conducted in vivo. Most gene-mutation tests involve in vitro single-cell systems, and only in vitro tests that are used most frequently are discussed here.

Gene mutations are usually single-base or base-pair alterations in DNA. Tests measuring those effects may be subdivided into tests that detect reverse or forward mutations. In general, the most sensitive tests for detecting gene mutations are the reverse-mutation methods, such as the bacterial reverse-mutation tests that use *Salmonella typhimurium* and *Escherichia coli* (EPA 1998r). The sensitivity of those organisms has been increased by modifications that allow chemicals to penetrate their cell walls more easily or prevent the DNA-repair process. The modifications have increased the ability of tests to detect mutagens. Some scientists view them as less relevant to the human situation and thus not very useful for providing quantitative information needed for risk assessment. However, these tests are extremely useful for hazard identification, and most scientists do not doubt the utility of single-cell gene-mutation systems for screening chemicals as the first steps in a tiered-testing approach. Positive test results are thought to indicate the potential muta-genicity of a chemical in both animals and humans. The *S. typhi-*

TABLE 2-2 Representative Genetic Toxicology Tests

Point or Gene Mutation	Chromosomal Aberration	DNA Damage
Salmonella microsome assay[a]	Human lymphocyte cells[e]	*E. coli* pol A (W3110/P3478)[i]
E. coli WP2uvrApKM101[b]	Chinese hamster ovary cells[d]	*Bacillus subtilis* rec (H17/M45)[i]
Mouse lymphoma L5178Y[c]	Mouse lymphoma L5178Y[c]	Rat liver primary cells[j]
Chinese hamster ovary cells[d]	Rodent bone marrow cells[f]	
	Dominant-lethal assay[g]	
	Heritable-translocation assay[h]	

[a]Maron and Ames 1983.
[b]EPA 1998r.
[c]Clive et al. 1979.
[d]Preston et al. 1981.
[e]Dean and Danford 1984.
[f]Kilian et al. 1977.
[g]Green et al. 1985.
[h]Generoso et al. 1980.
[i]Leifer et al. 1981.
[j]Williams 1977.

murium assay has also been shown to be useful for detecting mutagenic carcinogens.

Other in vitro systems that can be used to detect gene or point mutations are mouse lymphoma, Chinese hamster ovary, and Chinese hamster V79 cells (EPA 1998s). The mouse lymphoma test is also capable of detecting chromosomal aberrations. With most in vitro methods, metabolic-activation capabilities should be added to allow detection of chemicals that require activation to produce mutagenicity.

Chromosomal Aberrations

Chromosomal aberrations may be thought of as structural or numerical alterations of chromosomes. After exposure to a clastogen (an agent that causes structural alterations), breaks or gaps in the continuous structure of the chromosome can be viewed with a microscope. Clastogens are thought to cause major harm to DNA. Mammalian cells in culture—such as Chinese hamster fibroblasts, human or animal peripheral lymphocytes, and cells from whole animals—can be used (EPA 1998t). In animals, structural alterations can be observed in somatic cells, such as peripheral blood cells, and in germ (spermatogonial) cells. The value of assessing structural chromosomal damage is in the corroboration of an agent as a potential mutagen, that is, producing not only gene mutations but also chromosomal alterations. In addition, results of in vitro testing can be verified in vivo with the same end point (chromosomal aberrations), and this is not easily done in the case of gene mutations. It must be recognized, however, that not all chemicals that produce gene mutations produce structural chromosomal alterations (Zeiger 1998), and some human carcinogens, such as asbestos (Jaurand 1997), do not produce gene mutations but do cause chromosomal mutations.

Another type of structural alteration is formation of micronuclei. Micronuclei can be defined as cytoplasmic chromatin-containing bodies that are formed when acentric chromosomal fragments or whole chromosomes lag during anaphase and fail to become incorporated into daughter-cell nuclei during cell division (FDA 2000b). The mammalian red-blood-cell micronucleus test detects chromosomal fragments in bone marrow and other tissues. It can also detect numerical alterations (aneuploidy) (FDA 2000b; Hayashi et al. 2000).

Chromosomal aberrations in somatic cells have implications for carcinogenicity, whereas aberrations in germ cells imply transmission of effects to future generations. A number of whole-animal tests are used to detect chromosomal aberrations. The more prominent are the rat or mouse bone marrow cytogenetic assay (EPA 1998u), the rat or mouse dominant-lethal test (EPA 1998v), and the mouse heritable-translocation test (EPA 1998w). The dominant-lethal and heritable-translocation tests detect only chromosomal abnormalities that are produced in male germ cells. The bone marrow and dominant-lethal tests are usually used in the second tier of toxicity testing, and the heritable-translocation test in the third tier.

DNA Damage

Tests that detect DNA damage do not detect actual mutations. They yield indirect evidence of mutagenicity in that they detect an interaction with DNA that has affected the repair process. *E. coli* and *B. subtilis* (EPA 1998x) and such mammalian cells as liver cells are usually

used. In bacteria, cell survival differs between a strain that is capable of
DNA repair and one that is not. The mutagen should cause more cell
killing in the strain devoid of repair. In mammalian cells, a process
called unscheduled DNA synthesis measures the repair (synthesis) of
DNA at a period other than the usual S phase of the cell cycle (Herbold
et al. 2001). The S phase is the stage of the cell cycle in which normal
DNA synthesis occurs. Non-S-phase synthesis is thought to occur when
the cell is undergoing the excision repair that has been induced by the interaction of a chemical with DNA. Tritiated thymidine is incorporated
into the DNA of cells undergoing unscheduled DNA synthesis, and the
resulting radiolabeled nuclear grains are counted. A substantial increase
in the number of grains in the treated cells signifies an effect. The effect
is indirect (that is, it does not detect a mutagenic event) and is usually
viewed as indicating potential mutagenicity in mammalian cells. The
method is used primarily as a screening assay at the first level of a tiered
approach.

In Vitro Tests for Cellular Toxicity and Other End Points

Thousands of chemicals are synthesized or brought to market each
year, and conventional toxicity testing has not kept up with the pace of
development. The cost of animal testing is high; it may exceed several
million dollars per substance. In addition, there are concerns about animal welfare because testing uses a large number of animals. The development of in vitro model systems to evaluate the toxicity of chemicals
and drugs and potentially to reduce overall costs and animal use is re-
ceiving increased attention.

In vitro model systems, in general, have been used for two primary
purposes: to gain a better (perhaps mechanistic) understanding of
chemical-induced toxicity in animals and possibly humans and to serve
as rapid screening systems for the toxicologic evaluation of chemicals,
which may complement in vivo toxicity testing or may replace some in
vivo models if scientifically validated and accepted by regulatory agencies. Commonly used in vitro models for assessing chemical toxicity include perfused organ preparations, isolated tissue preparations, single-
cell suspensions, and cell-culture systems, such as primary cell cultures
and mammalian cell lines. Of these in vitro models, cell-culture systems
have been used more often by investigators because they are reliable, reproducible, and relatively inexpensive experimental systems to assess
chemical toxicity at the cellular level. In vitro tests that assess cellular

toxicity may be characterized as tests that measure cellular functions and tests that measure cell death. Functional assays typically measure reversible events that reflect a state of impairment or reversible cell injury or toxicity, and cell-death assays (often referred to as permeability assays) estimate the failure of the permeability barrier of the plasma membrane, which represents irreversible cell injury or loss of cell viability.

As described previously (Table 2-2), cultured human and other mammalian cells, such as lymphocytes and fibroblasts, are an integral part of EPA test guidelines for the determination of gene mutation, sister-chromatid exchange, chromosomal aberrations, and unscheduled DNA synthesis. EPA has accepted a battery of in vitro tests to evaluate the high-production-volume chemicals. In that battery, in vivo cytogenetics can be replaced by in vitro cytogenetics. In addition, nonanimal methods for assessing cellular toxicity have been used for many years by manufacturers to support inhouse decisions about product development before conducting conventional animal testing (Bruner et al. 1996). The recommendation of nonanimal methods for use in regulatory toxicology has occurred case by case since the development of the field of validation and the establishment of the Interagency Coordinating Committee on the Validation of Alternative Methods (ICCVAM) and its European counterpart, the European Centre for the Validation of Alternative Methods (ECVAM) in the 1990s. In recent years, ICCVAM and ECVAM have validated (or accepted as validated) and recommended for regulatory acceptance a number of in vitro tests. As seen in Table 2-3, the validated assays have primarily been assays that evaluate cytotoxicity or cell viability. The process for validating these methods has established their strengths and weaknesses and in some cases limited their applicability to specific chemical classes.

In addition to tests that have been validated and recommended for regulatory use, there are many examples in the literature of a wide variety of mammalian cells that are used to evaluate the toxicity and efficacy of chemical agents and to investigate mechanisms of toxicity. Several examples are discussed briefly below.

The Developmental Therapeutics Program at the National Cancer Institute (DTP 2005) developed an in vitro cell-line screen to support its drug-discovery program. The system, designed to screen up to 20,000 compounds per year for potential anticancer activity, uses 60 human tumor cell lines, representing leukemia, melanoma, and cancers of the

TABLE 2-3 Representative Validated In Vitro Methods for Cellular-Toxicity Testing[a]

Method	Purpose	Cells	End Points	Validated and Recommended for Regulatory Use by
Embryonic stem-cell test	Screening assay to identify potentially embryotoxic substances and to classify chemicals into three categories (strong, weak, and not embryotoxic)	Mouse embryonic stem (ES) cells and 3T3 adult mouse cell line	Inhibition of ES cell differentiation and inhibition of ES and 3T3 cell growth	ECVAM
Micromass test	Assay to evaluate ability of substances to inhibit differentiation	Micromass cultures of rat limb bud cells	Inhibition of differentiation, viability, and growth	ECVAM
Embryotoxicity testing in postimplantation rat whole-embryo culture	Assay to identify substances that induce malformation that results in embryo toxicity	Cultured rat embryos containing 1-5 somites	Embryo morphology after 48 hours of culture	ECVAM
In vitro 3T3 NRU phototoxicity test[b]	Assay to detect the phototoxicity induced by the combined action of a chemical and light	3T3 adult mouse fibroblast cell line	Neutral red uptake to indicate cytotoxicity	ECVAM

(Continued)

TABLE 2-3 *(Continued)*

Method	Purpose	Cells	End Points	Validated and Recommended for Regulatory Use by
EPIDERM skin-corrosivity test[c]	Assay to detect ability of a substance to cause skin corrosivity	EPIDERM human epidermal model system	Cell viability as determined by reduction of mitochondrial dehydrogenase activity measured by formazan production from MTT	ECVAM (replacement); ICCVAM (screen in tiered-testing strategy)
EPISKIN corrosivity test[c]	Assay to evaluate skin corrosivity	EPISKIN reconstructed human epidermis	Cell viability as determined by reduction of mitochondrial dehydrogenase activity measured by formazan production from MTT	ECVAM (replacement); ICCVAM (screen in tiered-testing strategy)
Rat skin transcutaneous test[d]	Assay to evaluate skin corrosivity	Skin disks prepared from rats 28-30 days old	Ability to produce a loss of normal stratum corneum integrity and barrier function, which is measured as a reduction of the inherent TER below a predetermined threshold level	ECVAM (replacement); ICCVAM (screen in tiered-testing strategy)

Corrositex[e]	Screening assay to predict skin corrosivity for classification and labeling	Reconstituted collagen membranes	Time required for a test material to pass through a biobarrier membrane (a reconstituted collagen matrix, constructed to have physiochemical properties similar to those of rat skin), and produce a visually detectable change	ECVAM (replacement) ICCVAM (screen in tired-testing strategy)
In vitro cytotoxicity test for assessing acute systemic toxicity[f]	Assay to select starting doses for acute oral toxicity in rodents	BALB/c3T3, normal human keratinocytes	Cell survival or viability as determined by the cell's ability to uptake and bind neutral red dye	ICCVAM (reduction)

[a]Information from the ECVAM Web site (ECVAM 2005) accessed September 2005 and personal communication with ICCVAM. The validations of the assays have established their strengths and weaknesses and in some cases limited their applicability to particular chemical classes or to levels within tiered-testing strategies. The Corrositex assay has been accepted for regulatory purposes and is approved for use by the Department of Transportation.
[b]OECD 2004d.
[c]OECD 2004e.
[d]OECD 2004f.
[e]OECD 2004g.
[f]Information from ICCVAM Web site (ICCVAM/NICEATM 2001) accessed October 2005.
Abbreviations: MTT, 3[4,5-dimethylthiazol-2-yl]-2,5-diphenyltetrazolium bromide; TER, transcutaneous electric resistance.

lung, colon, brain, ovary, breast, prostate, and kidney. The aim is to rank synthetic compounds or natural-product samples for selective growth inhibition or killing of particular tumor cell lines for further evaluation.

An engineered cell line, the MCL-5 human lymphoblast line, was designed to express five human cytochrome P450s (CYPs) (Crespi et al. 1991). The human CYPs incorporated into the MCL-5 cell line are the major P450s implicated in creating toxic metabolites and include CYPs 1A1, 1A2, 3A4, 2A6, and 2E1 and epoxide hydrolase. The MCL-5 assay an differentiate between parent and metabolite toxicity in one assay by comparing the concentrations of chemicals that produce 50% inhibition of growth (IC_{50}) in the MCL-5 cells with the IC_{50} values obtained in a control cell line cH2, which does not express the P450s.

Considerable effort has been devoted to the development of human skin or surface epithelium equivalents to replace animals in toxicity testing and in mechanism studies (Ponec 2002). The human tissue equivalents mimic native tissue to a high degree. They originate in human keratinocytes that are cultured on a variety of matrices and can be mixed with other cell types found in skin, including dendritic cells, melanocytes, and fibroblasts. Human three-dimensional tissue models for various epithelial tissues—including epidermal, corneal, esophageal, oral, tracheal and bronchial, ectocervical, and vaginal—have been developed and are commercially available. Those tissue equivalents form a typical multilayer epithelium and express markers of epithelial differentiation (Ponec 2002). Two such models are used in corrosivity assays that have been validated and approved for regulatory use by ECVAM and ICCVAM (see Table 2-3).

The greatest level of xenobiotic biotransformation is attributed to the liver, and considerable effort has been devoted to the development of various liver cell and tissue-culture systems for use in studies of xenobiotic metabolism and toxicity. A number of recent reviews (Guillouzo 1998; Lerche-Langrand and Toutain 2000; de Kanter et al. 2002; Groneberg et al. 2002; Gebhardt et al. 2003; Brandon et al. 2003) have covered the advantages and limitations of various in vitro liver cell systems, including isolated perfused liver, isolated hepatocytes in short-term suspension and primary monolayer culture, various liver cell lines, and liver precision-cut slice cultures. The isolated liver-perfusion system most closely resembles the in vivo liver in maintenance of tissue architecture and cell-cell interaction. Its use can reduce but not eliminate animal use. Those in vitro systems were initially developed with rodent liver in research laboratories investigating liver biochemistry, mechanisms of hepatocyte growth control, the role of cell-cell interactions, xenobiotic phase I and phase II biotransformation processes, and mechanisms of toxicity. They are widely used in drug development, where the objective is to obtain information that can be reliably extrapolated to the in vivo situation with regard to metabolism, drug-drug interactions, and mechanisms of toxicity. The increased availability of human liver (liver that is not usable for transplantation but is suitable for research studies) has provided data directly relevant to human liver biotransformation capacity and allowed comparison with data generated with rodent liver. Other organs and tissues also have biotransformation capacity that may contribute to tissue-specific toxicity, and the preparation of precision-cut tissue slices from various tissues (such as lung, kidney, and intestine) for use in studying tissue-specific biotransformation and toxicity has recently been reviewed by de Kanter et al. (2002).

Although in vitro systems are being used to assess the effects of chemicals on a number of cell types, only a few of the methods have been validated and recommended for acceptance by European and U.S. groups for regulatory purposes (see Table 2-3). Future research should be directed toward the refinement of existing methods and the development of new alternatives. As the level of technologic sophistication advances, an increased understanding of cellular mechanisms of toxicity will be achieved. Although in vitro model systems are now recognized as powerful tools in toxicity testing, their full potential remains largely undeveloped.

MECHANISTIC TESTING: METABOLISM AND PHARMACOKINETIC STUDIES[2]

Metabolism studies are used to identify potential biotransformation products after administration of test chemicals to experimental animals and to examine the potential for accumulation of compound or metabolites in tissues with repeated dosing. Pharmacokinetic (PK) studies can evaluate the time course of changes in concentrations in the body, persistence of radioactivity after dosing with radiolabeled compounds, or accumulation with multiple dosing. The EPA test guidelines for metabolism and pharmacokinetic studies (EPA 1998y) require a suite of studies

---

[2]Together, metabolism and pharmacokinetic studies have traditionally been referred to as ADME (absorption, distribution, metabolism, and excretion) studies.

to be submitted for review under FIFRA for pesticides. EPA requires development of data on metabolite formation, absorption, distribution, biotransformation, and excretion. The data are intended to aid in understanding mechanisms of toxicity and in determining whether animal toxicity studies are adequate for testing toxicity concerns associated with plant metabolites of the pesticide that might be present on raw agricultural products. The guidelines discuss two tiers of tests: a required core group of studies (tier 1) and a followup group (tier 2).

The tier 1 studies are more closely linked to specific requirements in the guidelines. In general, the guidelines expect tier 1 studies to satisfy regulatory requirements for biotransformation and PK data. Tier 1 tests use a radiolabeled test chemical to evaluate material balance, identify metabolites, and assess distribution and persistence of the radiolabel in various tissues. A single low-dose study is necessary with at least four young adult male animals, typically rats. The rationale for selection of dose route, specific dose rate, and animal sex needs to be described in the final report. By using radioactivity, the studies assess net behavior of the administered radiolabel rather than any specific characteristics of the parent chemical or important metabolites. The metabolism portion requires identification of all metabolites that constitute more than 5% of the original dose, from which the study report should propose a schematic of the pathways of metabolism.

Tier 2 studies should be designed to answer questions about chemical disposition on the basis of tier 1 results or other toxicity-test results. Any tier 2 study would be conducted according to agreement between a registrant and EPA, and considerable flexibility is allowable in tier 2 studies. Possible tier 2 studies identified in the test guidelines are evaluation of the extent of absorption, tissue-distribution time courses, plasma kinetics, and enzyme induction. The latter two still rely on radioactivity rather than speciation of the label or tracking of the parent chemical by chemical-specific analytic methods. The guidelines for tier 2 studies encourage physiologically based pharmacokinetic (PBPK) modeling and development of data needed for structured PK models.[3]

---

[3]PK models describe the time course in relation to model parameters that account for chemical movement and reactions within the body and elimination. In PBPK models, time-course behaviors are directly related to blood flows, tissue solubility, metabolic constants, and routes and processes of elimination. PBPK models have been developed to facilitate extrapolations required in human health risk assessments by predicting tissue doses of toxic compounds under a variety of exposure conditions in test animals and people, thereby allowing pre-

The guidelines also note that PBPK studies with parent chemicals might be submitted in lieu of other tier 1 studies if it is determined that PBPK studies would satisfy the metabolism and pharmacokinetics guidelines. Despite the encouragement, the guidelines do not define the studies required to generate datasets to support development of PK or PBPK models for any specific compound or class of compounds (that is, there are no formal guidelines for PK or PBPK model development). Current test guidelines for metabolism and PK studies will need to be revised to produce data more useful for PBPK model development. For example, the collection of data on radioactivity does not characterize the kinetics of specific forms of the compound (parent compound and major metabolites), and studies of individual compounds are needed for PBPK model development.

In general, the EPA guideline specifies a minimal dataset without providing a strong basis for evaluating alternative test strategies that might provide results more pertinent to human health risk assessment. The emphasis on measurement of radioactivity without speciation into specific compounds limits the value of the kinetic data collected in accordance with the guideline. Nevertheless, the studies are useful for assessing biotransformation products, for inferring the role of the parent chemical and metabolites in specific toxic responses, and for getting a sense of the overall pharmacokinetics of a compound in a test animal.

REFERENCES

Barton, H., K. Baetcke, J.E. Chambers, J. Diliberto, N.G. Doerrer, J.H. Driver, C.H. Hastings, S. Iyengar, R. Krieger, T. Pastoor, B. Stahl, and C. Timchalk. 2005. Draft ADME White Paper: The Acquisition and Application of Absorption, Distribution, Metabolism, and Excretion (ADME) Data in Agricultural Chemical Safety Assessments. International Life Sciences Institute, Health and Environmental Sciences Institute, Washington, DC. 65 pp [online]. Available: http://www.hesiglobal.org/NR/rdonlyres/E53BA06D-2E40-43A3-9845-2A8218DDE7D3/0/ADMEDraft PaperJan05.pdf [accessed Nov. 16, 2005].

Brandon, E.F., C.D. Raap, I. Meijerman, J.H. Beijnen, and J.H. Schellens. 2003. An update on in vitro test methods in human hepatic drug biotransformation research: Pros and cons. Toxicol. Appl. Pharmacol. 189(3):233-246.

dictions of tissue doses in humans on the basis of interspecies differences in physiology and metabolism (Clewell 1995a,b).

Bruner, L.H., G.J. Carr, M. Chamberlain, and R.D. Curren. 1996. Validation of alternative methods for toxicity testing. Toxicol. In Vitro 10(4):479-501.

Brusick, D. 2001. Genetic toxicology. Pp. 819-852 in Principals and Methods of Toxicology, 4th Ed, A.W. Hayes, ed. Philadelphia, PA: Taylor and Francis.

Clewell, III, H.J. 1995a. Incorporating biological information in quantitative risk assessment: An example with methylene chloride. Toxicology 102 (1-2):83-94.

Clewell, III, H.J. 1995b. The application of physiologically based pharmacokinetic modeling in human health risk assessment of hazardous substances. Toxicol. Lett. 79(1-3):207-217.

Clive, D., K.O. Johnson, J.F. Spector, A.G. Batson, and M.M. Brown. 1979. Validation and characterization of the L5178Y/TK+/- mouse lymphoma mutagen assay system. Mutat. Res. 59(1):61-108.

Cohen, S.M., D. Robinson, and J. MacDonald. 2001. Alternative models for carcinogenicity testing. Toxicol. Sci. 64(1):14-19.

Cory-Slechta, D.A., K.M. Crofton, J.A. Foran, J.F. Ross, L.P. Sheets, B. Weiss, and B. Mileson. 2001. Methods to identify and characterize developmental neurotoxicity for human health risk assessment. I: Behavioral effects. Environ. Health Perspect. 109(Suppl.1):79-91.

Crespi, C.L., F.J. Gonzalez, D.T. Steimel, T.R. Turner, H.V. Gelboin, B.W. Penman, and R. Langenbach. 1991. A metabolically competent human cell line expressing five cDNAs encoding procarcinogen-activating enzymes: Application to mutagen testing. Chem. Res. Toxicol. 4(5):566-572.

Dean, B.J., and N. Danford. 1984. Assays for the detection of chemically-induced chromosome damage in cultured mammalian cells. Pp. 187-232 in Mutagenicity Testing: A Practical Approach, S. Venitt, and J.M. Parry, eds. Oxford: IRL Press.

de Kanter, R., M. Monshouwer, D.K. Meijer, and G.M. Groothuis. 2002. Precision-cut organ slices as a tool to study toxicity and metabolism of xenobiotics with special reference to non-hepatic tissues. Curr. Drug Metab. 3(1):39-59.

Dorman, D.C., S.L. Allen, J.Z. Byczkowski, L. Claudio, J.E. Fisher, Jr., J.W. Fisher, G.J. Harry, A.A. Li, S.L. Makris, S. Padilla, L.G. Sultatos, and B.E. Mileson. 2001. Methods to identify and characterize developmental neurotoxicity for human health risk assessment. III: Pharmacokinetic and pharmacodynamic considerations. Environ. Health Perspect. 109 (Suppl.1):101-111.

DTP (Developmental Therapeutics Program). 2005. DTP Human Tumor Cell Line Screen. Screening, Developmental Therapeutics Program, National Cancer Institute [online]. Available: http://www.dtp.nci.nih.gov/branches/btb/ivclsp.html [accessed March 18, 2005].

ECVAM (European Centre for the Validation Alternative Methods). 2005. Scientifically Validated Methods [online]. Available: http://ecvam.jrc.cec. eu.int/index.htm [accessed March 16, 2005].

EPA (U.S. Environmental Protection Agency). 1996. Microbial Pesticide Test Guidelines OPPTS 885.3000. Background-Mammalian Toxicity/ Pathogenicity/Infectivity. EPA 712-C-96-314. Office of Prevention, Pesticides, and Toxic Substances, U.S. Environmental Protection Agency, Washington, DC [online]. Available: http://www.epa.gov/opptsfrs/ OPPTS_Harmonized/885_Microbial_Pesticide_Test_Guidelines/Series/88 5-3000.pdf [accessed March 14, 2005].

EPA (U.S. Environmental Protection Agency). 1998a. Health Effects Test Guidelines OPPTS 870.1000. Acute Toxicity Testing-Background. EPA 712-C-98-189. Office of Prevention, Pesticides, and Toxic Substances, U.S. Environmental Protection Agency, Washington, DC [online]. Available: http://www.epa.gov/opptsfrs/OPPTS_Harmonized/870_Health_ Effects_Test_Guidelines/Series/870-1000.pdf [accessed October 7, 2005].

EPA (U.S. Environmental Protection Agency). 1998b. Health Effects Test Guidelines OPPTS 870.1100. Acute Oral Toxicity. EPA 712-C-98-190. Office of Prevention, Pesticides, and Toxic Substances, U.S. Environmental Protection Agency, Washington, DC [online]. Available: http:// www.epa.gov/opptsfrs/OPPTS_Harmonized/870_Health_Effects_Test_Gu idelines/Series/870-1100.pdf [accessed March 15, 2005].

EPA (U.S. Environmental Protection Agency). 1998c. Health Effects Test Guidelines OPPTS 870.1300. Acute Inhalation Toxicity. EPA-712-C-98-193. Office of Prevention, Pesticides, and Toxic Substances, U.S. Environmental Protection Agency, Washington, DC [online]. Available: http://www.epa.gov/opptsfrs/OPPTS_Harmonized/870_Health_Effects_ Test_Guidelines/Series/870-1300.pdf [accessed October 7, 2005].

EPA (U.S. Environmental Protection Agency). 1998d. Health Effects Test Guidelines OPPTS 870.1200. Acute Dermal Toxicity. EPA 712-C-98-192. Office of Prevention, Pesticides, and Toxic Substances, U.S. Environmental Protection Agency, Washington, DC [online]. Available: http://www.epa.gov/opptsfrs/OPPTS_Harmonized/870_Health_Effects_ Test_Guidelines/Series/870-200.pdf [accessed October 7, 2005].

EPA (U.S. Environmental Protection Agency). 1998e. Health Effects Test Guidelines OPPTS 870.4100. Chronic Toxicity. EPA 712-C-98-210. Office of Prevention, Pesticides, and Toxic Substances, U.S. Environmental Protection Agency, Washington, DC [online]. Available: http://www.epa. gov/opptsfrs/OPPTS_Harmonized/870_Health_Effects_Test_Guidelines/S eries/870-4100.pdf [accessed March 15, 2005].

EPA (U.S. Environmental Protection Agency). 1998f. Health Effects Test Guidelines OPPTS 870.4200. Carcinogenicity. EPA 712-C-98-211. Office of Prevention, Pesticides, and Toxic Substances, U.S. Environmental Protection Agency, Washington, DC [online]. Available: http://www.epa.

gov/opptsfrs/OPPTS_Harmonized/870_Health_Effects_Test_Guidelines/S
eries/870-4200.pdf [accessed March 15, 2005].

EPA (U.S. Environmental Protection Agency). 1998g. Health Effects Test
Guidelines   OPPTS   870.4300.   Combined   Chronic   Toxicity/
Carcinogenicity. EPA 712-C-98-212. Office of Prevention, Pesticides,
and Toxic Substances, U.S. Environmental Protection Agency, Washington, DC [online]. Available: http://www.epa.gov/opptsfrs/OPPTS_
Harmonized/870_Health_Effects_Test_Guidelines/Series/870-4300.pdf
[accessed March 15, 2005].

EPA (U.S. Environmental Protection Agency). 1998h. Health Effects Test
Guidelines OPPTS 870.3800. Reproduction and Fertility Effects. EPA
712-C-98-208. Office of Prevention, Pesticides, and Toxic Substances,
U.S. Environmental Protection Agency, Washington, DC [online]. Avail-
able: http://www.epa.gov/opptsfrs/OPPTS_Harmonized/870_Health_
Effects_Test_Guidelines/Series/870-3800.pdf [accessed March 15, 2005].

EPA (U.S. Environmental Protection Agency). 1998i. Guidelines for Neurotoxicity Assessment. EPA/630/R-95/001F. National Center for Environmental Assessment, U.S. Environmental Protection Agency, Washington,
DC [online]. Available: http://www.epa.gov/ncea/raf/pdfs/neurotox.pdf
[accessed March 15, 2005].

EPA (U.S. Environmental Protection Agency). 1998j. Health Effects Test
Guidelines OPPTS 870.3700. Prenatal Developmental Toxicity Study.
EPA 712-C-98-207. Office of Prevention, Pesticides, and Toxic Sub-
stances, U.S. Environmental Protection Agency, Washington, DC [online].
Available: http://www.epa.gov/opptsfrs/OPPTS_Harmonized/870_Health
_Effects_Test_Guidelines/Series/870-3700.pdf   [accessed   March   15,
2005].

EPA (U.S. Environmental Protection Agency). 1998k. Health Effects Test
Guidelines OPPTS 870.6200. Neurotoxicity Screening Battery. EPA
712-C-98-238. Office of Prevention, Pesticides, and Toxic Substances,
U.S. Environmental Protection Agency, Washington, DC [online]. Avail-
able: http://www.epa.gov/opptsfrs/OPPTS_Harmonized/870_Health_
Effects_Test_Guidelines/Series/870-6200.pdf [accessed March 15, 2005].

EPA (U.S. Environmental Protection Agency). 1998l. Health Effects Test
Guidelines OPPTS 870.6300. Developmental Neurotoxicity Study. EPA
712-C-98-239. Office of Prevention, Pesticides, and Toxic Substances,
U.S. Environmental Protection Agency, Washington, DC [online]. Avail-
able: http://www.epa.gov/opptsfrs/OPPTS_Harmonized/870_Health_
Effects_Test_Guidelines/Series/870-6300.pdf [accessed March 15, 2005].

EPA (U.S. Environmental Protection Agency). 1998m. Health Effects Test
Guidelines OPPTS 870.6100. Acute and 28-Day Delayed Neurotoxicity
of Organophosphorus Substances. EPA 712-C-98-237. Office of Prevention, Pesticides, and Toxic Substances, U.S. Environmental Protection
Agency, Washington, DC [online]. Available: http://www.epa.gov/

opptsfrs/OPPTS_Harmonized/870_Health_Effects_Test_Guidelines/Series /870-6100.pdf [accessed March 15, 2005].

EPA (U.S. Environmental Protection Agency). 1998n. Health Effects Test Guidelines OPPTS 870.6500. Schedule-Controlled Operant Behavior. EPA 712-C-98-240. Office of Prevention, Pesticides, and Toxic Substances, U.S. Environmental Protection Agency, Washington, DC [online]. Available: http://www.epa.gov/opptsfrs/OPPTS_Harmonized/870_Health _Effects_Test_Guidelines/Series/870-6500.pdf [accessed March 15, 2005].

EPA (U.S. Environmental Protection Agency). 1998o. Health Effects Test Guidelines OPPTS 870.6850. Peripheral Nerve Function. EPA 712-C-98-241. Office of Prevention, Pesticides, and Toxic Substances, U.S. Environmental Protection Agency, Washington, DC [online]. Available: http://www.epa.gov/opptsfrs/OPPTS_Harmonized/870_Health_Effects_ Test_Guidelines/Series/870-6850.pdf [accessed March 15, 2005].

EPA (U.S. Environmental Protection Agency). 1998p. Health Effects Test Guidelines OPPTS 870.6855. Neurophysiology: Sensory Evoked Potentials. EPA 712-C-98-242. Office of Prevention, Pesticides, and Toxic Substances, U.S. Environmental Protection Agency, Washington, DC [online]. Available: http://www.epa.gov/opptsfrs/OPPTS_Harmonized/ 870_Health_Effects_Test_Guidelines/Series/870-6855.pdf [accessed March 15, 2005].

EPA (U.S. Environmental Protection Agency). 1998q. Health Effects Test Guidelines OPPTS 870.7800. Immunotoxicity. EPA 712-C-98-351. Office of Prevention, Pesticides, and Toxic Substances, U.S. Environmental Protection Agency, Washington, DC [online]. Available: http://www. epa.gov/opptsfrs/OPPTS_Harmonized/870_Health_Effects_Test_Guidelin es/Series/870-7800.pdf [accessed March 9, 2005].

EPA (U.S. Environmental Protection Agency). 1998r. Health Effects Test Guidelines OPPTS 870.5100. Bacterial Reverse Mutation Test. EPA 712-C-98-247. Office of Prevention, Pesticides, and Toxic Substances, U.S. Environmental Protection Agency, Washington, DC [online]. Available: http://www.epa.gov/opptsfrs/OPPTS_Harmonized/870_Health_ Effects_Test_Guidelines/Series/870-5100.pdf [accessed February 18, 2005].

EPA (U.S. Environmental Protection Agency). 1998s. Health Effects Test Guidelines OPPTS 870.5300. In Vitro Mammalian Cell Gene Mutation Test. EPA 712-C-98-221. Office of Prevention, Pesticides, and Toxic Substances, U.S. Environmental Protection Agency, Washington, DC [online]. Available: http://www.epa.gov/opptsfrs/OPPTS_Harmonized/ 870_Health_Effects_Test_Guidelines/Series/870-5300.pdf [accessed February 18, 2005].

EPA (U.S. Environmental Protection Agency). 1998t. Health Effects Test Guidelines OPPTS 870.5380. Mammalian Spermatogonial Chromosome

Aberration Test. EPA 712-C-98-224. Office of Prevention, Pesticides, and Toxic Substances, U.S. Environmental Protection Agency, Washington, DC [online]. Available: http://www.epa.gov/opptsfrs/OPPTS_Harmonized/870_Health_Effects_Test_Guidelines/Series/870-5380.pdf [accessed March 14, 2005].

EPA (U.S. Environmental Protection Agency). 1998u. Health Effects Test Guidelines OPPTS 870.5385. Mammalian Bone Marrow Chromosome Aberration Test. EPA 712-C-98-225. Office of Prevention, Pesticides, and Toxic Substances, U.S. Environmental Protection Agency, Washington, DC [online]. Available: http://www.epa.gov/opptsfrs/OPPTS_Harmonized/870_Health_Effects_Test_Guidelines/Series/870-5385.pdf [accessed February 18, 2005].

EPA (U.S. Environmental Protection Agency). 1998v. Health Effects Test Guidelines OPPTS 870.5450. Rodent Dominant Lethal Assay. EPA 712-C-98-227. Office of Prevention, Pesticides, and Toxic Substances, U.S. Environmental Protection Agency, Washington, DC [online]. Available: http://www.epa.gov/opptsfrs/OPPTS_Harmonized/870_Health_Effects_Test_Guidelines/Series/870-5450.pdf [accessed February 18, 2005].

EPA (U.S. Environmental Protection Agency). 1998w. Health Effects Test Guidelines OPPTS 870.5460. Rodent Heritable Translocation Assays. EPA 712-C-98-228. Office of Prevention, Pesticides, and Toxic Substances, U.S. Environmental Protection Agency, Washington, DC [online]. Available: http://www.epa.gov/opptsfrs/OPPTS_Harmonized/870_Health_Effects_Test_Guidelines/Series/870-5460.pdf [accessed February 18, 2005].

EPA (U.S. Environmental Protection Agency). 1998x. Health Effects Test Guidelines OPPTS 870.5500. Bacterial DNA Damage or Repair Tests. EPA 712-C-98-229. Office of Prevention, Pesticides, and Toxic Substances, U.S. Environmental Protection Agency, Washington, DC [online]. Available: http://www.epa.gov/opptsfrs/OPPTS_Harmonized/870_Health_Effects_Test_Guidelines/Series/870-5500.pdf [accessed February 18, 2005].

EPA (U.S. Environmental Protection Agency). 1998y. Health Effects Test Guidelines OPPTS 870.7485. Metabolism and Pharmacokinetics. EPA 712–C–98–244. Office of Prevention, Pesticides, and Toxic Substances, U.S. Environmental Protection Agency, Washington, DC [online]. Available: http://www.epa.gov/opptsfrs/OPPTS_Harmonized/870_Health_Effects_Test_Guidelines/Series/870-7485.pdf [accessed June 15, 2005].

EPA (Environmental Protection Agency). 1999. Toxicology Data Requirements for Assessing Risks of Pesticide Exposure to Children's Health. Draft Report of the Toxicology Working Group of the 10x Task Force, U.S. Environmental Protection Agency. April 28, 1999 [online]. Available: http://www.epa.gov/scipoly/sap/1999/may/10xtx428.pdf [accessed October 24, 2005].

EPA (U.S. Environmental Protection Agency). 2003. Health Effects Test Guidelines OPPTS 870.2600. Skin Sensitization. EPA 712-C-03-197. Office of Prevention, Pesticides, and Toxic Substances, U.S. Environmental Protection Agency, Washington, DC [online]. Available: http://iccvam.niehs.nih.gov/docs/EPA/870r_2600.pdf [accessed March 16, 2005].

FDA (U.S. Food and Drug Administration). 2000a. Short-term tests for genetic toxicity. Section IV.C.1 in Redbook 2000, Toxicological Principles for the Safety Assessment of Food Ingredients, Office of Food Additive Safety, U.S. Food and Drug Administration [online]. Available: http://www.cfsan.fda.gov/~redbook/red-ivc1.html [accessed March 14, 2005].

FDA (U.S. Food and Drug Administration). 2000b. Mammalian erythrocyte micronucleus test. Section IV.C.1.d. in Redbook 2000, Toxicological Principles for the Safety Assessment of Food Ingredients, Office of Food Additive Safety, U.S. Food and Drug Administration [online]. Available: http://www.cfsan.fda.gov/~redbook/redivc1d.html [accessed February 18, 2005].

FDA (U.S. Food and Drug Administration). 2002. Guidance for Industry: Immunotoxicology Evaluation of Investigational New Drugs. Center for Drug Evaluation and Research, Food and Drug Administration, U.S. Department of Health and Human Services [online]. Available: http://www.fda.gov/cder/guidance/4945fnl.PDF [accessed March 16, 2005].

Flamm, W.G., W. d'Aguanno, L. Fishbein, S. Green, H.V. Malling, V. Mayer, M. Prival, G. Wolff, and E. Zeiger. 1977. Approaches to determining the mutagenic properties of chemicals: Risk to future generations. J. Environ. Pathol. Toxicol. 1(2):301-352.

Garman, R.H., A.S. Fix, B.S. Jortner, K.F. Jensen, J.F. Hardisty, L. Claudio, and S. Ferenc. 2001. Methods to identify and characterize developmental neurotoxicity for human health risk assessment. II: Neuropathology. Environ. Health Perspect. 109(Suppl. 1):93-100.

Gebhardt, R., J.G. Hengstler, D. Muller, R. Glockner, P. Buenning, B. Laube, E. Schmelzer, M. Ullrich, D. Utesch, N. Hewitt, M. Ringel, B.R. Hilz, A. Bader, A. Langsch, T. Koose, H.J. Burger, J. Maas, and F. Oesch. 2003. New hepatocyte in vitro systems for drug metabolism: Metabolic capacity and recommendations for application in basic research and drug development, standard operation procedures. Drug Metab. Rev. 35(2-3):145-213.

Generoso, W.M., J.B. Bishop, D.G. Gosslee, G.W. Newell, C.J. Sheu, and E. von Halle. 1980. Heritable translocation test in mice. Mutat. Res. 76(2): 191-215.

Green, S., A. Auletta, J. Fabricant, R. Kapp, M. Manandhar, C.J. Sheu, J. Springer, and B. Whitfield. 1985. Current status of bioassays in genetic toxicology—the dominant lethal assay. A report of the U.S. Enviromental Protection Agency Gene-Tox Program. Mutat Res. 154(1):49-67.

Groneberg, D.A., C. Grosse-Siestrup, and A. Fischer. 2002. In vitro models to study hepatotoxicity. Toxicol. Pathol. 30(3):394-399.

Guillouzo, A. 1998. Liver cell models in vitro toxicology. Environ. Health Perspect. 106(Suppl. 2):511-532.

Hayashi, M., J.T. MacGregor, D.G. Gatehouse, I.D. Adler, D.H. Blakey, S.D. Dertinger, G. Krishna, T. Morita, A. Russo, and S. Sutou. 2000. In vivo rodent erythrocyte micronucleus assay. II. Some aspects of protocol design including repeated treatments, integration with toxicity testing, and automated scoring. Environ. Mol. Mutagen. 35(3):234-252.

Herbold, B.A., S.Y. Brendler-Schwaab, and H.J. Ahr. 2001. Ciprofloxacin: In vivo genotoxicity studies. Mutat. Res. 498(1-2):193-205.

ICCVAM/NICEATM (Interagency Coordinating Committee on the Validation of Alternative Methods and National Toxicology Program Interagency Center for the Evaluation of Alternative Toxicological Methods). 2001. Report of the International Workshop on In Vitro Methods for Assessing Acute Systematic Toxicity. NIH 04-4499. National Institutes of Health, U.S. Public Health Service, Department of Health and Human Services [online]. Available: http://iccvam.niehs.nih.gov/methods/invidocs/finalall.pdf [accessed Dec. 7, 2005].

Jaurand, M.C. 1997. Mechanisms of fiber-induced genotoxicity. Environ. Health Perspect. 105(Suppl. 5):1073-1084.

Kilian, D.J., F.M. Moreland, M.C. Benge, M.S. Legator, and E.B. Whorton, Jr. 1977. A collaborative cytogenetics study to measure and minimize interlaboratory variation. Mutat Res. 44(1):97-104.

Leifer, Z., T. Kada, M. Mandel, E. Zeiger, R. Stafford, and H.S. Rosenkranz. 1981. An evaluation of tests using DNA repair-deficient bacteria for predicting genotoxicity and carcinogenicity. A report of the U.S. EPA's Gene-Tox. Program. Mutat. Res. 87(3):211-297.

Lerche-Langrand, C., and H.J. Toutain. 2000. Precision-cut liver slices: Characteristics and use for in vitro pharmaco-toxicology. Toxicology 153(1-3):221-253.

Li, A.A. 2005. Regulatory developmental neurotoxicology testing: Data evaluation for risk assessment purposes. Environ. Toxicol. Pharmacol. 19(3):727-733.

Luster, M.I., C. Portier, D.G. Pait, K.L. White Jr., C. Gennings, A.E. Munson, and G.J. Rosenthal. 1992. Risk assessment in immunotoxicology. I. Sensitivity and predictability of immune tests. Fundam. Appl. Toxicol. 18(2):200-210.

Luster, M.I., C. Portier, D.G. Pait, G.J. Rosenthal, D.R. Germolec, E. Corsini, B.L. Blaylock, P. Pollock, Y. Kouchi, W. Craig, K.L. White, A.E. Munson, and C.E. Comment. 1993. Risk assessment in immunotoxicology. II. Relationships between immune and host resistance tests. Fundam. Appl. Toxicol. 21(1):71-82.

Makris, S., K. Raffaelle, W. Sette, and J. Seed. 1998. A Retrospective Analysis of Twelve Developmental Neurotoxicity Studies Submitted to the U.S. Environmental Protection Agency Office of Prevention, Pesticides and Toxic Substances (OPPTS), Washington, DC. Draft Report, Nov.12, 1998 [online]. Available: http://www.epa.gov/scipoly/sap/1998/december/neuro.pdf [accessed November 7, 2005].

Maron, D.M., and B.N. Ames. 1983. Revised methods for the Salmonella mutagenicity test. Mutat Res. 113(3-4):173-215.

Middaugh, L.D., D. Dow-Edwards, A.A. Li, J.D. Sandler, J. Seed, L.P. Sheets, D.L. Shuey, W. Slikker Jr., W.P. Weisenburger, L.D. Wise, and M.R. Selwyn. 2003. Neurobehavioral assessment: A survey of use and value in safety assessment studies. Toxicol. Sci. 76(2): 250-261.

Moser, V.C., S.E. Bowen, A.A. Li, W.S. Sette, and W.P. Weisenburger. 2000. Cognitive evaluation: Is it needed in neurotoxicity screening? Symposium presented at the annual Behavioral Toxicology Society meeting, May 1999. Neurotoxicol. Teratol. 22(6):785-798.

NRC (National Research Council). 1992. Environmental Neurotoxicology. Washington, DC: National Academy Press.

NTP (National Toxicology Program). 1999. The Murine Local Lymph Node Assay: A Test Method for Assessing the Allergic Contact Dermatitis Potential of Chemicals/Compounds. NIH Publication No. 99-4494. National Toxicology Program, Research Triangle Park, NC. February 1999 [online]. Available: http://iccvam.niehs.nih.gov/methods/llnadocs/llnarep.pdf [accessed March 16, 2005].

OECD (Organisation for Economic Cooperation and Development). 1981. Acute Inhalation Toxicity. Chemicals Testing Guidelines No. 403. Paris: Organisation for Economic Cooperation and Development. May 12, 1981.

OECD (Organisation for Economic Cooperation and Development). 1983. One-Generation Reproduction Toxicity Study. Chemicals Testing Guidelines No. 415. Paris: Organisation for Economic Cooperation and Development. May 26, 1983.

OECD (Organisation for Economic Cooperation and Development). 1987. Acute Dermal Toxicity. Chemicals Testing Guidelines No. 402. Paris: Organisation for Economic Cooperation and Development. February 24, 1987.

OECD (Organisation for Economic Cooperation and Development). 1995a. Reproduction/Developmental Toxicity Screening Test. Chemicals Testing Guidelines No. 421. Paris: Organisation for Economic Cooperation and Development. July 27, 1995.

OECD (Organisation for Economic Cooperation and Development). 1995b. Repeated Dose 28-Day Oral Toxicity Study in Rodents. Chemicals Testing Guidelines No. 407. Paris: Organisation for Economic Cooperation and Development. July 27, 1995.

OECD (Organisation for Economic Cooperation and Development). 1996. Combined Repeated Dose Toxicity Study With the Reproduction/ Developmental Toxicity Screening Test. Chemicals Testing Guidelines No. 422. Paris: Organisation for Economic Cooperation and Development. March 22, 1996.

OECD (Organisation for Economic Cooperation and Development). 1997. Neurotoxicity Study in Rodents. Chemicals Testing Guidelines No. 424. Paris: Organisation for Economic Cooperation and Development. July 21, 1997.

OECD (Organisation for Economic Cooperation and Development). 1998. Repeated Dose 90-Day Oral Toxicity Study in Rodents. Chemicals Testing Guidelines No. 408. Paris: Organisation for Economic Cooperation and Development September 21, 1998.

OECD (Organisation for Economic Cooperation and Development). 2001a. Harmonized Integrated Classification System for Human Health and Environmental Hazards of Chemical Substances and Mixtures. ENV/JM/ MONO(2001)6. OECD Series on Testing and Assessment No. 33. Organisation for Economic Cooperation and Development [online]. Available: http://iccvam.niehs.nih.gov/docs/OECD/hclfinaw.pdf [accessed March 14, 2005].

OECD (Organisation for Economic Cooperation and Development). 2001b. Prenatal Developmental Study. Chemicals Testing Guidelines No. 414. Paris: Organisation for Economic Cooperation and Development. January 22, 2001.

OECD (Organisation for Economic Cooperation and Development). 2001c. Two-Generation Reproduction Toxicity Study. Chemicals Testing Guidelines No. 416. Paris: Organisation for Economic Cooperation and Development. January 22, 2001.

OECD (Organisation for Economic Cooperation and Development). 2001d. Acute Oral Toxicity-Fixed Dose Method. Chemicals Testing Guidelines No. 420. Paris: Organisation for Economic Cooperation and Development. December 20, 2001.

OECD (Organisation for Economic Cooperation and Development). 2001e. Acute Oral Toxicity-Acute Toxic Class Method. Chemicals Testing Guidelines No. 423. Paris: Organisation for Economic Cooperation and Development. December 20, 2001.

OECD (Organisation for Economic Cooperation and Development). 2001f. Acute Oral Toxicity: Up and Down Procedure. Chemicals Testing Guidelines No. 425. Paris: Organisation for Economic Cooperation and Development. December 20, 2001.

OECD (Organisation for Economic Cooperation and Development). 2004a. OECD Guidelines for the Testing of Chemicals, 15th Addendum. Paris: Organisation for Economic Cooperation and Development. April 2004.

OECD (Organisation for Economic Cooperation and Development). 2004b. Draft Guidance Document on Reproductive Toxicity Testing and Assessment, Series on Testing and Assessment No. 43. Environment Directorate, Organisation for Economic Cooperation and Development, November 10, 2004 [online]. Available: http://www.oecd.org/dataoecd/38/46/34030071.pdf [accessed March 15, 2005].

OECD (Organisation for Economic Cooperation and Development). 2004c. Guidance Document for Neurotoxicity Testing. ENV/JM/MONO (2004)25. Series on Testing and Assessment No. 20. Environment Directorate, Organisation for Economic Cooperation and Development, November 24, 2004 [online]. Available: http://appli1.oecd.org/olis/2004doc.nsf/43bb6130e5e86e5fc12569fa005d004c/c631c2551f372e47c12 56f580058e823/$FILE/JT00174673.PDF [accessed Nov. 16, 2005].

OECD (Organisation for Economic Cooperation and Development). 2004d. In vitro 3T3 NRU phototoxicity test. Chemicals Testing Guidelines No. 432. Paris: Organisation for Economic Cooperation and Development.

OECD (Organisation for Economic Cooperation and Development). 2004e. In vitro skin corrosion: Human skin model test. Chemicals Testing Guidelines No. 431. Paris: Organisation for Economic Cooperation and Development.

OECD (Organisation for Economic Cooperation and Development). 2004f. In vitro skin corrosion: Transcutaneious electrical resistance test (TER). Chemicals Testing Guidelines No. 430. Paris: Organisation for Economic Cooperation and Development. Available: http://www.oecd.org/dataoecd/63/23/32037794.pdf [accessed Dec. 7, 2005].

OECD (Organisation for Economic Cooperation and Development). 2004g. In vitro membrane barrier test method for skin corrosion. Draft guideline. Chemicals Testing Guidelines No. 435. Paris: Organisation for Economic Cooperation and Development [online]. Available: http://www.oecd.org/dataoecd/63/23/32037794.pdf [accessed Dec. 7, 2005].

OECD (Organisation for Economic Cooperation and Development). 2005. Chemicals Testing: OECD Guidelines for the Testing of Chemicals - Sections 4: Health Effects [online]. Available: http://www.oecd.org/document/55/0,2340,en_2649_34377_2349687_1_1_1_1,00.html [accessed March 14, 2005].

Ponec, M. 2002. Skin constructs for replacement of skin tissues for in vitro testing. Adv. Drug Deliv. Rev. 54(Suppl.1):S19-S30.

Preston, R.J., W. Au, M.A. Bender, J.G. Brewen, A.V. Carrano, J.A. Heddle, A.F. McFee, S. Wolff, and J.S. Wassom. 1981. Mammalian in vivo and in vitro cytogenetic assays: A report of the U.S. EPA's Gene-Tox Program. Mutat. Res. 87(2):143-188.

Raffaele, K.C., W.F. Sette, S.L. Makris, V.C. Moser, and K.M. Crofton. 2003. Motor activity in developmental neurotoxicity testing: A cross-laboratory comparison of control data. The Toxicologist 72(Suppl. 1):123[598].

Raffaele, K.C., M. Gilbert, K.M. Crofton, S.L. Makris, and W.F. Sette. 2004. Learning and memory tests in developmental neurotoxicity testing: A cross-laboratory comparison of control data. The Toxicologist 78(Suppl. 1):276[1342].

Rice, D.C., and S. Barone. 2000. Critical periods of vulnerability for the developing nervous system: Evidence from humans and animal models. Environ. Health Perspect. 108(Suppl. 3):511-533.

Sette, W.F., K.M. Crofton, S.L. Makris, J. Doherty, and K.C. Raffaele. 2004. Auditory startle reflex habituation in developmental neurotoxicity testing: A cross-laboratory comparison of control data.   The Toxicologist 78(Suppl. 1):275[1341].

WHO (World Health Organization). 2001. Environmental Health Committee 223: Neurotoxicity Risk Assessment for Human Health: Principles and Approaches [online]. Available: http://www.inchem.org/documents/ehc/ehc/ehc223.htm [accessed October 7, 2005].

Williams, G.M. 1977. Detection of chemical carcinogens by unscheduled DNA synthesis in rat liver primary cell cultures. Cancer Res. 37(6):1845-1851.

Zeiger, E. 1998. Identification of rodent carcinogens and noncarcinogens using genetic toxicity tests: Premises, promises and performance. Regul. Toxicol. Pharmacol. 28(2):85-95.

3

Human Data

Human data generally are not a part of toxicity-testing strategies despite the importance of human responses to potentially toxic agents. Although animal toxicity studies and in vitro studies provide relevant information on potential adverse health effects of exposure to an agent, they can miss an effect relevant to the human population. As mentioned in Chapter 2, a famous example is thalidomide, to which rats are highly resistant but human fetuses are exquisitely sensitive. Studying the human population also provides an opportunity to evaluate the effects of the full variety of agents in the complex contexts of workplaces and daily lives. The large populations also provide an enormous sample size in which rare effects might be detected. If the human experience is not evaluated, animal-to-human extrapolations are tenuous (for example, some responses observed in animals may not be relevant to the human population). Human data provide a benchmark for those extrapolations.

Given the importance of human data, this chapter reviews the various types of human data, provides examples of the use of human data in regulatory analyses, and considers the challenges to and possible advances in studies of the human population. Regarding availability of human data, clearly, no population data will be available on a chemical newly introduced to the marketplace, although there may be controlled-exposure data, such as those from a clinical trial conducted on a pharmaceutical. Population data will be available only on chemicals that have been in production for some time, perhaps several decades. Thus, differences in data availability on new versus existing chemicals should be considered in developing the role of human data in any toxicity-testing strategy.

CLINICAL OR CONTROLLED-EXPOSURE STUDIES

Humans are often intentionally exposed to various agents to evaluate possible health effects. Most intentional exposures occur during the development of potential new medicines and are used to characterize their efficacy and safety. The clinical-trial process is subject to numerous regulations, guidance, monitoring, and reporting obligations that attach primary importance to patient well-being. Specifically, the trials are conducted under multiple federal regulations (21 CFR 21, 50, 54, 312 [2004]), good-clinical-practice guidelines, and technical requirements established by the International Conference on Harmonization (EMEA 2002). Such guidelines and regulatory requirements determine regulatory oversight processes, trial conduct, ethical review, informed consent, monitoring of drug supplies, adverse-event monitoring, and data integrity and quality assurance.

Preclinical safety testing of investigational new drugs must satisfy the appropriate regulatory bodies that the first clinical trials in humans will pose minimal risk for subjects. The exhaustive nature of the preclinical assessment, which includes high-dose acute and multidose chronic studies in animals, means that only a few potential new drugs will be deemed sufficiently safe for administration to human volunteers. The trial process itself is separated into distinct phases, and the study protocol for each phase is subject to review by an institutional review board or ethics committee. The phases of clinical trials are as follows:

- *Phase 1.* This stage, typically performed in fewer than 100 healthy volunteers, is designed to establish dose-range tolerance. It may include a carefully controlled and monitored dose-escalation protocol. For some disease indications, such as cancer and HIV, the Food and Drug Administration supports an accelerated process in which efficacy and tolerance are assessed simultaneously in patients with the disease in question.
- *Phase 2.* For most indications, this stage is designed to refine dose ranges, establishing efficacy and safety in typically 100-500 selected patients who represent the target population. Drug tolerance is monitored. Parallel safety studies in animals are also run to characterize potential adverse effects that may be a consequence of high-dose or long-term exposure and to characterize specific end points, such as reproductive and developmental effects.

- *Phase 3.* These studies, which are usually multicenter and possibly multinational, involve thousands of representative patients and enable assessment of the efficacy and safety of the proposed new drug at doses characterized in the previous phases. Successful completion of this stage, with demonstrated efficacy and manageable side effects, is necessary for approval to market the drug.
- *Phase 4.* Postmarketing surveillance, which may include further clinical trials, involves the collection of further data on drug efficacy and safety in the broader patient population.

Some environmental agents, such as ozone and perchlorate, have been studied with controlled exposures of volunteers. Those studies have provided information on pharmacokinetics and pharmacodynamics at environmentally relevant concentrations. The ethical implications of such studies have been raised, and guidance on their conduct and their use for regulatory purposes is being debated (NRC 2004a).

CASE REPORTS

Many human toxicants were first recognized by astute clinicians who reported their suspicions that the occurrence of some rare disease in association with an unusual exposure was more than coincidence. Case reports include detailed medical information that has been collected on a single patient or a series of similar patients. Clinical information may be gathered from private physicians, hospitals, clinics, and ambulatory-care facilities to investigate and understand disease etiology. Some clinical information on acute exposures may be obtained from the Toxic Exposure Surveillance System, which contains data on over 36 million human poison-exposure cases compiled by the American Association of Poison Control Centers (Watson et al. 2004).

Case reports are particularly useful in investigating exposures on which there is little or no reported human toxicity information. Some of the most informative case reports are derived from occupational settings. Workers in industrial settings are often the first to show adverse effects of an agent because of their high or chronic exposure. One example is the recognition of the causative link between vinyl chloride and hepatic angiosarcoma among polyvinyl chloride workers. Zymbal gland carcinomas, nephroblastomas, and hepatic angiosarcomas were observed in rats exposed to vinyl chloride by inhalation in August 1972. In December 1973, a case of malignant hepatic angiosarcoma in a polyvinyl chloride production worker was associated with occupational vinyl chloride exposure (Creech and Johnson 1974; Maltoni et al. 1981). Retrospective occupational cohort studies later confirmed the connection. Thus, researchers identified a new cause (exposure to vinyl chloride monomer) of a rare disease previously associated only with medical use of Thorotrast and occupational exposure to inorganic arsenic (Falk et al. 1981).

EPIDEMIOLOGIC STUDIES

Epidemiologic studies typically investigate the relationship between exposure to a substance and potential health effects in a human population. There are several study designs and different approaches to organizing and classifying them. Table 3-1 provides one perspective on defining epidemiologic study designs and their basic characteristics.

TABLE 3-1 Examples of Epidemiologic Study Designs

Study Design	Characteristics	Comments
Cohort	A study in which the individual is the unit of observation. A cohort (large group of people) is defined and evaluated over a particular period to determine the occurrence of a health-related outcome and its possible relationship to a given exposure. *Prospective cohort studies* monitor a disease-free population selected at the beginning of the study for the occurrence of health effects associated with a given exposure. *Retrospective cohort studies* evaluate a cohort after the outcome has occurred, and exposure information is estimated on the basis of historical records, subjects' memories, or job descriptions.	The primary purpose of a cohort study is to establish the incidence or occurrence of new cases of the health outcome among the exposed and unexposed groups to estimate a relative risk between two groups. An unbiased cohort study can reflect the cause-effect temporal sequence of events with regard to an exposure and an outcome.

Case-Control	A study that involves the recruitment of a series of cases with a specific disease and a series of disease-free controls. Comparisons of exposure to the agent of interest are then made between the cases and control series.	Higher exposure to a specific agent among the cases than among the controls suggests that the risk of the disease of interest may be increased as a consequence of exposure.
Cross-Sectional	A study in which disease prevalence and exposures are evaluated in a cohort at a single time (that is, people are not followed over time).	An important distinction between cohort and cross-sectional studies is that a cohort study selects an at-risk population, but a cross-sectional study selects people who are then classified as having or not having the disease on the basis of information collected after selection.
Ecologic	A study that examines exposure and risk factors on a group level (generally studies of geographically defined populations). *Cross-sectional ecologic studies* compare aggregate exposures and outcomes in communities in the same period. *Time-trend ecologic studies* compare aggregate exposures and outcomes in the same community over time.	An association observed between two variables on an aggregate level does not necessarily represent an association on an individual level (known as the ecologic fallacy). Causation cannot be established by such studies, but they can supply useful supporting information.

Epidemiologic studies are often referred to as occupational or environmental depending on whether the study population is exposed in the workplace or through daily living, respectively. Because of higher exposures in some workplaces relative to the general environment, the occupational setting has provided valuable information on the potential adverse effects of various chemicals. Although occupational exposure monitoring is done primarily for purposes of industrial hygiene and compliance with occupational exposure guidelines, the resulting data are often useful in occupational epidemiologic studies. Large industries may also have disease surveillance programs, which can be used to provide health-outcome data for occupational studies. However, occupational data may be biased because of the healthy-worker effect, which is defined as a population bias resulting from reduced recruitment or early withdrawal of less-healthy persons from the worker population. That bias diminishes the possibility of observing a significant increase in risk. Other factors that may affect the reliability of occupational, as well as environmental, studies are poor or no control for confounding factors, nonrandom sampling of study subjects, exposure and disease measurement error, and missing data due to subject nonresponse or losses to followup. Regardless of the possible study limitations, occupational and environmental epidemiologic studies can provide a systematic evaluation of human exposures and possible outcomes, are valuable in the risk-assessment process, and are relevant in determining the adequacy of regulatory standards for chemicals that are already widely used.

USE OF HUMAN DATA FOR REGULATORY ANALYSES

Human data have been used to estimate risk and to establish standards for environmental and occupational exposures. Table 3-2 provides

TABLE 3-2 Examples of Risks or Standards Derived from Human Data For Drinking-Water Standards and Advisories

Arsenic	Cancer risks estimated from studies of bladder and other cancers in populations consuming arsenic-contaminated water (EPA 1984; NRC 1999; OEHHA 2004a)
Benzene	Cancer risks estimated from studies of leukemia in workers in the Pliofilm industry in the United States and a large cohort of workers from various industries in China (EPA 1998; OEHHA 2001)
Nitrate	Reference levels estimated from studies of infants exposed to nitrate at >20 mg of nitrate-nitrogen per liter in drinking water used to prepare their formula (Bosch et al. 1950; Walton 1951)
Perchlorate	Reference levels estimated from controlled-exposure studies of inhibition of thyroid iodide uptake in perchlorate-exposed humans (OEHHA 2004b; NRC 2005)

For Air-Pollutant Standards and Advisories	
Arsenic	Cancer risks estimated from studies of lung cancer in workers in the smelter and pesticide manufacturing industries (Roth 1958; Ott et al. 1974; Tokudome and Kuratsune 1976; Rencher et al. 1977; Axelson et al. 1978; Mabuchi et al. 1979; Matanoski et al. 1981; Enterline and Marsh 1982; Lee-Feldstein 1983)
Benzene	Cancer risks estimated from studies of leukemia in workers in the Pliofilm industry (Rinsky et al. 1981, 1987)
Cadmium	Cancer risks estimated from studies of lung cancer in workers in the cadmium smelter industry (Thun et al. 1985)
Diesel exhaust	Cancer risks estimated from studies of lung cancer in rail workers (Crump 2001; OEHHA/ALA 2001)
Hexavalent chromium	Cancer risks estimated from studies of lung cancer in workers in the chromate production industry (Mancuso and Hueper 1951; Mancuso 1975, 1997)
Ozone	Criteria standard derived from controlled human chamber studies of lung-function decrements and respiratory symptoms after ozone exposure (EPA 2005a) or epidemiologic studies of premature mortality, respiratory hospitalization, and asthma exacerbation (OEHHA 2004c)
Particulate Matter	Criteria standard derived from studies of correlations of premature mortality and fine-particle exposure in various U.S. cities (EPA 2005b)
Vinyl chloride	Risks estimated from studies of hepatic angiosarcoma and other cancers in workers in the U.S. polyvinyl chloride industry (Feron et al. 1981)

For Food Residues	
Aflatoxin	Widely recognized as a known human carcinogen on the basis of numerous studies of populations in China and Africa consuming contaminated foods (IARC 1993; FDA 2003); risks estimated from data in studies of Chinese populations controlled for confounding by hepatitis infection (Wu-Williams et al. 1992)
Methyl-mercury	Reference exposure levels initially established from studies of poisoning incident in Iraq where people consumed grain treated with organomercurial pesticides and currently established from studies of associations of developmental effects and hair concentration in populations consuming large amounts of seafood (NRC 2000; EPA 2001)

examples of the use of human data to set water, air, and food standards or advisories. In March 2004, the Environmental Protection Agency (EPA) Risk Assessment Task Force reviewed a sample of EPA's Integrated Risk Information System database to estimate how often human data were used in developing reference concentrations (RfCs), reference doses (RfDs), or cancer risk assessments (EPA 2004). Of the 15 RfC determinations reviewed, eight included human data, and four of the eight used the human data as the principal basis for determining the RfCs. Of the 42 RfD determinations reviewed, nine included human data, and five of the nine used the human data as the principal basis to derive the RfDs. Of the 27 classifications of carcinogenicity reviewed, 10 identified human data, and four of the 10 used the human data to make the classification of carcinogenicity. When human data were available but not used as the principal data, a variety of reasons were provided, including the questionable relevance of the exposures, concurrent exposure to other chemicals, imprecise measurements of exposure and duration, inadequate consideration of confounding factors, inadequate statistical power, insufficient time after exposure to observe outcome, and the difficulty of using null results from epidemiologic studies.

Epidemiologic studies have played a particularly important role in the assessment of population health risks associated with air pollutants. Two kinds of epidemiologic studies have shown that pollutants in ambient air are associated with adverse health outcomes (Cohen et al. 2003; Samet and Krewski 2005). Adverse health effects of short-term exposures to air pollutants have been consistently demonstrated in studies that relate daily fluctuations in pollutant exposure to hospital admissions, mortality (Samet et al. 2000), and perinatal health outcomes (Liu et al. 2003). Long-term cohort mortality studies, most notably the Harvard six-cities study (Dockery et al. 1993) and the American Cancer Society (ACS) study (Pope et al. 1995), have shown that long-term exposure to particulate air pollution is associated with increased cardiopulmonary and possibly lung-cancer mortality (Pope et al. 2002). Recent analyses of the ACS cohort conducted by Pope et al. (2003) have suggested that cardiovascular mortality associated with particulate air pollution is consistent with pathophysiologic mechanisms of accelerated atherosclerosis. That hypothesis is supported by toxicologic data suggesting that particulate air pollution may lead to the induction of endothelins and cytokines, which may in turn lead to atherosclerosis (NRC 2004b).

CHALLENGES TO THE ADVANCEMENT OF EPIDEMIOLOGY

Epidemiologic studies have been widely criticized on the grounds that their methodologic limitations make it difficult to draw clear associations between exposure and disease. Those limitations have made it difficult to use epidemiologic data in regulatory risk assessments. Three of the most common problems are that only uncertain or indirect estimates of human exposure are available; that epidemiologic studies may identify associations with chemical classes, such as organophosphate pesticides, or with consumer-product categories, such as insecticides, rather than specific chemicals; and that the indeterminate and often long latency period between exposure and disease creates logistical challenges for study design and adds to the uncertainty of results. Because of the complexity of epidemiologic datasets, there is a need to develop and refine statistical methods of analysis to address critical data issues, such as random and systematic exposure-measurement error, selection bias, the effects of residual confounding and unmeasured covariates, and errors in health-outcome ascertainment.

Good exposure assessment is critical for population-based research in environmental health to reduce the likelihood of biased results and to provide information that is valid and useful for informing public-health decision-making. Adequate assessment of human exposure to an environmental agent includes determining the exposure intensity, frequency, and duration. A good exposure assessment should answer several questions: Are people exposed to the environmental agent? If so, what is the statistical distribution of exposures in the population? How do exposures depend on personal characteristics, such as age, place of residence, and work in a particular area of a factory? How have exposures changed over time? Through what pathways are people exposed? Evaluation of exposures of children and evaluation of exposures to mixtures are other exposure issues. Children often have exposure pathways that differ from those in adults because of their propensity for hand-to-mouth activities, and children may be exposed to a higher dose relative to their body weight than adults in the same setting. Mixtures typically are not addressed in exposure assessment; when exposures to mixtures are measured, incorporation of the results into a risk assessment is often impossible because of the lack of information about whether the combined exposures act in an additive, less than additive, antagonistic, or synergistic manner.

Despite all the complex issues, human studies are needed to determine actual exposures; laboratory investigations cannot do that. Most epidemiologic studies have estimated human exposure to suspected environmental toxicants by collecting questionnaire data on past behavior patterns related to exposure. Some studies have used job title or place of residence to categorize the exposure of study subjects. Those methods often result in simplistic exposure categorizations, such as exposed and unexposed, which are of limited use in quantitative risk assessment. Furthermore, the errors in such data can lead to misclassification of exposure, which can increase variances, introduce bias, or both. Nondifferential random exposure misclassification will bias studies toward a null result, and a study may fail to detect or adequately measure a true association. In contrast, systematic exposure-measurement error can lead to bias toward or away from a null result.

Exposure assessment has improved in recent years. Environmental monitoring has provided useful data for exposure assessments. For example, testing foods and drinking water for contaminants has allowed scientists to create reasonable exposure estimates from the test results and data on food and water consumption patterns. Exposure modeling also has proved helpful in exposure assessment. Models can be constructed to estimate exposures to chemicals in food, water, and air and from various household scenarios, such as a toddler playing on a lawn or carpet. Exposure models gradually improve as they are tested against monitored data and can be useful for generating exposure assessments for regulatory risk assessments. Some models are not publicly available for scientific scrutiny and so cannot be assessed for validity. Such models should not be used in the development of regulations or advisories until they have been shown to be valid and reliable. The emerging fields that hold much promise for improving exposure assessment and other issues mentioned are discussed briefly below.

CONTRIBUTIONS OF EMERGING
FIELDS TO EPIDEMIOLOGY

Improving the science of epidemiology so that it can improve the effectiveness of toxicity testing of environmental agents should have high priority. New fields are emerging that may help to overcome the issues discussed. Specifically, developments in biomonitoring, molecular and genetic epidemiology, and environmental health tracking hold great potential for overcoming some of the major historical limitations of epidemiology and are discussed in the following sections.

Biomonitoring

Biomonitoring is the measurement of biomarkers in blood, urine, and tissues. A biomarker is defined as "any substance, structure or process that can be measured in the body or its products and influence or predict the incidence of outcome or disease" (WHO 2001). Biomarkers should ideally be both specific to a particular environmental agent and sufficiently sensitive to reflect the effects of low-level exposure to that agent. The incorporation of biomarkers in epidemiologic research offers considerable potential to improve exposure estimates and detection of adverse health effects of environmental agents in population-based studies. In applying biomarkers of exposure, pharmacokinetic models are needed to define the relationship between exposure to a compound and the concentration of that compound or its metabolites in body tissues.

Biomarkers of Exposure

Biomarkers of exposure—such as lead in blood or deciduous teeth and polychlorinated biphenyls in blood or breast milk—have been in wide use for many years and have been critical in creating a large and robust epidemiologic database on a variety of toxicants. The Centers for Disease Control and Prevention (CDC) *National Report on Human Exposure to Environmental Chemicals* provides a continuing assessment of the U.S. population's exposure to environmental chemicals based on a statistical sample of the general population. The first report (CDC 2001) presented biomonitoring data on 27 chemicals; the second report (CDC 2003) on 116 chemicals, including the original 27; and the third report (CDC 2005) on 149 chemicals, including the 116 from the second report. CDC's work is one example of estimating population exposures with biomarkers of exposure. Examples of innovative uses of biomarkers of exposure in population studies include researchers sampling participants for residues of relevant contaminants (Nordstrom et al. 2000; Pavuk et al. 2003). By decreasing exposure misclassification, the studies have helped to overcome one of the major hurdles of epidemiology. However, biomonitoring does not fully overcome dose uncertainty and is applicable mostly to chemicals with long biologic half-lives and to disease end points with relatively short latency periods. Some researchers have successfully used biomarkers of exposure for relatively short-lived toxicants in prospective cohort studies (Whyatt et al. 2004; Murray et al. 2004). Such an approach can allow clearer demarcation of the exposure status of members of the cohort, although it does not necessarily allow extrapolation to dose.

Biomarkers of exposure are critically important for strengthening epidemiologic research, but they can be used for other purposes. Population-based exposure surveys can be used to determine the distribution of exposure to specific toxicants, to identify groups with high exposures, and to track trends. Such data can be useful in setting priorities when there is a need to determine whether exposure is widespread and in the risk-assessment process when there is a need to identify highly exposed populations. Agencies can also use biomarkers of exposure to track the effectiveness of regulatory efforts or identify a need for regulatory attention.

Biomarkers of Effect

Epidemiologic studies of such outcomes as cancer and chronic disease are particularly difficult because years or even decades may elapse between an exposure and the manifestations of symptomatic disease. In some cases, the mechanism of action of an environmental toxicant is sufficiently well understood that biomarkers of effect have been developed and used in human health risk assessment. For example, perchlorate, a drinking-water contaminant, is known to inhibit the uptake of iodide by the thyroid and thus possibly decrease the production of thyroid hormones. Inhibition of radioiodide uptake measured in a human clinical study has been used as a biomarker of effect in risk assessments performed by state environmental agencies and by the National Research Council (OEHHA 2004b; NRC 2005). Markers of airway inflammation (such as nitric oxide in exhaled breath; inflammatory cells, cytokines, and chemokines in bronchoalveolar-lavage fluid; and RANTES gene activation) have been used in studies of the effects of ozone, diesel exhaust, and other air pollutants on rodents and humans (Pandya et al. 2002). Clinical tests of effect, such as forced expiratory volume in 1 sec (FEV_1), have been used in risk assessments that have formed the basis of regulation of pollutants, including ozone (Gauderman et al. 2000).

Some biomarkers appear to be markers of both exposure and effect. For example, polycyclic aromatic hydrocarbon (PAH) DNA adducts in biologic samples can be used to assess exposure to PAHs, evaluate the potential for an early event in a multistep process of carcinogenesis, and perhaps even predict cancer risk in some groups (Peluso et al. 2005). Such biomarkers might be useful in risk assessment in which prevention of an early effect would protect human health.

Molecular and Genetic Epidemiology

Molecular epidemiology and genetic epidemiology identify molecular biomarkers of exposure and effect and incorporate them into study designs to investigate gene-environment interactions and their associations with the etiology and distribution of disease. Studies incorporating biomarkers have demonstrated that genetic consequences of human exposure are measurable and definable in tissues from exposed people (Schroeder et al. 2003; Perera et al. 2005). For example, spontaneous chromosomal aberrations detected in peripheral blood lymphocytes have been shown to identify humans at increased cancer risk (Bonassi et al. 2000; Chien et al. 2004; Shao et al. 2004). The use of other biomarkers, such as DNA adducts and urinary hydroxypyrene, in epidemiologic studies has shown that exposed groups have considerable increases in DNA-associated damage or excreted metabolites (Wiencke et al. 1995; Siwinska et al. 2004; Peluso et al. 2005). Those studies have been useful in helping to understand the relative risks associated with different routes of exposure (for example, dermal vs inhalation) and in providing evidence to support the mechanistic understanding of relationships between exposure, disease, and potential modifiers of absorbed dose (Schurdak and Randerath 1989; Turteltaub et al. 1993; McClean et al. 2004).

Delineation of genetic variation is also proving important in defining potential differences in susceptibility to environmental toxicants. Genetic variation is well known to give rise to heritable disease states, and recent work suggests that common normal genetic polymorphisms may in some cases be associated with an increase in susceptibility to toxicants (Caporaso and Goldstein 1997; Singh 2003). It is critical to remember that once a genetic polymorphism has been identified, it is not a simple task to determine its mechanism of action. For example, it is not always clear whether a particular variant itself has a biologically distinct action or if it is linked to another variation that is functionally important. Phenotypic characterization of genotypic variation is critical to the application of such data in population studies.

Environmental-Health Tracking

New efforts to collect data relevant to environmental health in human populations systematically may hold promise for improving the quality and quantity of data available for epidemiologic studies. The Institute of Medicine has stated that "every public health agency [should] regularly and systematically collect, assemble, analyze, and make available information on the health of the community, including statistics on health status, community health needs, and epidemiologic and other studies of health problems" (IOM 1988).

That recommendation has been implemented for some types of diseases but poorly developed for others. Many infectious diseases, such as rabies and influenza, are intensively tracked in the United States to facilitate public-health responses. Birth defects and cancer are tracked in some states, and the data are centrally compiled at CDC and the National Cancer Institute, respectively. However, hospital-discharge data and medical-billing data, which are sometimes useful for developing disease patterns, are not centralized, are of mixed quality, and are not useful for many chronic diseases.

Because most diseases are multifactorial, elucidation of the environmental causes of human disease requires data on exposure to environmental agents that can be linked to specific adverse health outcomes. However, few systems at the state or national level track many of the exposures and health effects that may be related to environmental hazards. The existing tracking systems are usually not compatible with each other, and data linkage is extremely difficult.

Over the last 5 years, there has been an effort to create a nationwide environmental public-health tracking (EPHT) program in up to 20 states and local regions with a coordinating center at CDC. The national EPHT program was established in 2002 with low funding, and its future is in some doubt. EPHT is defined as the "ongoing collection, integration, analysis, and interpretation of data about environmental hazards, exposure to environmental hazards, and human health effects potentially related to exposure to environmental hazards" (CDC 2004). An integrated EPHT system includes three components: hazard tracking, exposure tracking, and disease tracking. The components are designed to be maintained in electronic files that can be linked to facilitate hypothesis generation and research.

There are at least four reasons to create an integrated environmental health surveillance system: tracking of environmental hazards, exposures, and disease can help to identify areas or groups in which exposure to an environmental hazard may be excessive and require reduction; trends can help to evaluate the success of environmental-protection and public-health measures; linkage of environmental-hazard information and disease information can help to generate hypotheses that require investigation; and a tracking network provides the foundation that researchers need to do scientific studies to identify the causes of disease.

REFERENCES

Axelson, O., E. Dahlgren, C.D. Jansson, and S.O. Rehnlund. 1978. Arsenic exposure and mortality: A case referent study from a Swedish copper smelter. Br. J. Ind. Med. 35(1):8-15.

Bonassi, S., L. Hagmar, U. Stromberg, A.H. Montagud, H. Tinnerberg, A. Forni, P. Heikkila, S. Wanders, P. Wilhardt, I.L. Hansteen, L.E. Knudsen, and H. Norppa. 2000. Chromosome aberrations in lymphocytes predict human cancer independently of exposure to carcinogens. Cancer Res. 60(6): 1619-1625.

Bosch, H.M., A.B. Rosefield, R. Huston, H.R. Shipman, and F.L. Woodward. 1950. Methemoglobinemia and Minnesota well supplies. J. Am. Water Works Assoc. 42:161-170.

Caporaso, N., and A. Goldstein. 1997. Issues involving biomarkers in the study of the genetics of human cancer. Pp. 237-250 in Application of Biomarkers in Cancer Epidemiology, P. Toniolo, P. Boffetta, D. Shuker, N. Rothman, B. Hulka, and N. Pearce, eds. IARC Science Publications No. 142. Lyon, France: IARC.

CDC (Centers for Disease Control and Prevention). 2001. National Report on Human Exposure to Environmental Chemicals. U.S. Department of Health and Human Services, Centers for Disease Control and Prevention, Atlanta, GA [online]. Available: http://www.noharm.org/details.cfm?ID =745&type=document [accessed March 25, 2005].

CDC (Centers for Disease Control and Prevention). 2003. Second National Report on Human Exposure to Environmental Chemicals. U.S. Department of Health and Human Services, Centers for Disease Control and Prevention, Atlanta, GA [online]. Available: http://www.serafin.ch/ toxicreport.pdf [accessed October 25, 2005].

CDC (Centers for Disease Control and Prevention). 2004. Environmental Public Health Tracking Program: Closing America's Environmental Public Health Gap 2004. U.S. Department of Health and Human Services, Center for Disease Control and Prevention, National Center for Environmental Health, Atlanta, GA [online]. Available: http://www.cdc.gov/nceh/tracking/aag04.htm [accessed March 22, 2005].

CDC (Centers for Disease Control and Prevention). 2005. Third National Report on Human Exposure to Environmental Chemicals. U.S. Department of Health and Human Services, Centers for Disease Control and Prevention, Atlanta, GA [online]. Available: http://www.cdc.gov/exposurereport/pdf/third_report_chemicals.pdf [accessed March 25, 2005].

Chien, C.W., M.C. Chiang, I.C. Ho, and T.C. Lee. 2004. Association of chromosomal alterations with arsenite-induced tumorigenicity of human HaCaT keratinocytes in nude mice. Environ Health Perspect. 112(17):1704-1710.

Cohen, A.J., D. Krewski, J. Samet, and R. Willes, eds. 2003. Health and Air Quality: Interpreting Science for Decision Makers. J. Toxicol. Environ. Health A 66(16-19):1489-1903.

Creech, J.L., and M.N. Johnson. 1974. Angiosarcoma of liver in the manufacture of polyvinyl chloride. J. Occup. Med. 16(3):150-151.

Crump, K. 2001. Modeling lung cancer risk from diesel exhaust: Suitability of the railroad worker cohort for quantitative risk assessment. Risk Anal. 21(1):19-23.

Dockery, D.W., C.A. Pope, III, X. Xu, J.D. Spengler, J.H. Ware, M.E. Fay, B.G. Ferris, Jr., and F.E. Speizer. 1993. An association between air pollution and mortality in six U.S. cities. N. Engl. J. Med. 329(24):1753-1759.

EMEA (European Agency for the Evaluation of Medicinal Products). 2002. Guideline for Good Clinical Practice, ICH Topic E 6, Step 5, Consolidated Guideline 1.5.96. The European Agency for the Evaluation of Medicinal Products, London, UK [online]. Available: http://www.emea.eu.int/pdfs/human/ich/013595en.pdf [accessed March 25, 2005].

Enterline, P.E., and G.M. Marsh. 1982. Cancer among workers exposed to arsenic and other substances in a copper smelter. Am. J. Epidemiol. 116(6): 895-911.

EPA (U.S. Environmental Protection Agency). 1984. Health Assessment Document for Inorganic Arsenic. Final Report. EPA 600/8-83/021F. U.S. Environmental Protection Agency, Office of Research and Development, Office of Health and Environmental Assessment, Environmental Criteria and Assessment Office, Research Triangle Park, NC.

EPA (U.S. Environmental Protection Agency). 1998. Carcinogenic Effects of Benzene: An Update. EPA/600/P-97/001F. National Center for Environmental Assessment, Office of Research and Development, U.S. Environmental Protection Agency, Washington, DC [online]. Available: http://www.epa.gov/ncea/pdfs/benzenef.pdf [accessed March 24, 2005].

EPA (U.S. Environmental Protection Agency). 2001. Methylmercury Criteria Document: Fish Tissue Criterion for Methylmercury to Protect Human Health Document. EPA-823-R-01-001. Office of Water, U.S. Environmental Protection Agency, Washington, DC [online]. Available: http://www.epa.gov/waterscience/criteria/methylmercury/document.html [accessed March 29, 2005].

EPA (Environmental Protection Agency). 2004. Pp. 72-82 in An Examination of EPA Risk Assessment Principles and Practice, Staff Paper Prepared for the U.S. Environmental Protection Agency by Members of the Risk Assessment Task Force. EPA/100/B-04/001. Office of the Science Advisor, U.S. Environmental Protection Agency, Washington, DC [online]. Available: http://www.epa.gov/OSA/ratf-final.pdf [accessed March 22, 2005].

EPA (Environmental Protection Agency). 2005a. Air Quality Criteria for Ozone and Related Photochemical Oxidants (First External Review Draft). EPA/600/R-05/004aA-cA. National Center for Environmental Assessment, U.S. Environmental Protection Agency, Washington, DC [online]. Available: http://cfpub.epa.gov/ncea/cfm/recordisplay.cfm?deid=114523 [accessed March 29, 2005].

EPA (Environmental Protection Agency). 2005b. Review of the National Ambient Air Quality Standards for Particulate Matter: Policy Assessment of Scientific and Technical Information. OAQPS Staff Paper-Second Draft. January 2005 [online]. Available: http://www.epa.gov/ttn/naaqs/standards/pm/data/pm_staff_paper_2nddraft.pdf [accessed March 29, 2005].

Falk, H., J. Herbert, S. Crowley, K.G. Ishak, L.B. Thomas, H. Popper, and G.G. Caldwell. 1981. Epidemiology of hepatic angiosarcoma in the United States: 1964-1974. Environ. Health Perspect. 41:107-113.

FDA (U.S. Food and Drug Administration). 2003. Natural Toxins: Aflatoxins. Foodborne Pathogenic Microorganisms and Natural Toxins Handbook: The Bud Bug Book. U.S. Food and Drug Administration, Center for Food Safety and Applied Nutrition [online]. Available: http://www.cfsan.fda.gov/~mow/chap41.html [accessed March 29, 2005].

Feron, V.J., C.F.M. Hendriksen, A.J. Speek, H.P. Til, and B.J. Spit. 1981. Life-span oral toxicity study of vinyl-chloride in rats. Food Cosmet. Toxicol. 19(3):317-333.

Gauderman, W.J., R. McConnell, F. Gilliland, S. London, D. Thomas, E. Avol, H. Vora, K. Berhane, E.B. Rappaport, F. Lurmann, H.G. Margolis, and J. Peters. 2000. Association between air pollution and lung function growth in southern California children. Am. J. Respir. Crit. Care Med. 162(4 Pt 1):1383-1390.

IARC (International Agency for Research on Cancer). 1993. Aflatoxins. Pp. 245-395 in Some Naturally Occurring Substances: Food Items and Constituents, Heterocyclic Aromatic Amines and Mycotoxins. IARC Monographs on the Evaluation of Carcinogenic Risks to Humans No. 56. Lyon, France: IARC.

IOM (Institute of Medicine). 1988. The Future of Public Health. Washington, DC: National Academy Press.

Lee-Feldstein, A. 1983. Arsenic and respiratory cancer in man: Follow-up of an occupational study. Pp. 245-254 in Arsenic: Industrial, Biomedical, and Environmental Perspectives, W. Lederer, and R. Fensterheim, eds. New York: Van Nostrand Reinhold.

Liu, S., D. Krewski, Y. Shi, Y. Chen, and R.T. Burnett. 2003. Association between gaseous ambient air pollutants and adverse pregnancy outcomes in Vancouver, Canada. Environ. Health Perspect. 111(14):1773-1778.

Mabuchi, K., A.M. Lilienfeld, and L. Snell. 1979. Lung cancer among pesticide workers exposed to inorganic arsenicals. Arch. Environ. Health 34(5):312-320.

Maltoni, C., G. Lefemine, A. Ciliberti, G. Coti, and D. Carretti. 1981. Carcinogenicity bioassays of vinyl chloride monomer: A model of risk assessment on an experimental basis. Environ. Health Perspect. 41:3-29.

Mancuso, T.F. 1975. Consideration of chromium as an industrial carcinogen. Pp. 343-356 in Proceedings of the International Conference on Heavy Metals in the Environment, October 27-31, 1975, Toronto, Ontario, Canada, T.C. Hutchinson, ed. Toronto: Institute for Environmental Studies.

Mancuso, T.F. 1997. Chromium as an industrial carcinogen: Part 1. Am. J. Ind. Med. 31(2):129-139.

Mancuso, T.F., and W.C. Hueper. 1951. Occupational cancer and other health hazards in a chromate plant: A medical appraisal. I. Lung cancers in chromate workers. Ind. Med. Surg. 20(8):358-363.

Matanoski, G., E. Landau, J. Tonascia, C. Lazar, E.A. Elliot, W. McEnroe, and K. King. 1981. Cancer mortality in an industrial area of Baltimore. Environ. Res. 25(1):8-28.

McClean, M.D., R.D. Rinehart, L. Ngo, E.A. Eisen, K.T. Kelsey, and R.F. Herrick. 2004. Inhalation and dermal exposure among asphalt paving workers. Ann. Occup. Hyg. 48(8):663-671.

Murray, C.S., A. Woodcock, F.I. Smillie, G. Cain, P. Kissen, and A. Custovic. 2004. Tobacco smoke exposure, wheeze, and atopy. Pediatr. Pulmonol. 37(6):492-498.

Nordstrom, M., L. Hardell, G. Lindstrom, H. Wingfors, K. Hardell, and A. Linde. 2000. Concentrations of organochlorines related to titers to Epstein-Barr virus early antigen IgG as risk factors for hairy cell leukemia. Environ. Health Perspect. 108(5):441-445.

NRC (National Research Council). 1999. Arsenic in Drinking Water. Washington, DC: National Academy Press.

NRC (National Research Council). 2000. Toxicological Effects of Methylmercury. Washington, DC: National Academy Press.

NRC (National Research Council). 2004a. Acute Exposure Guideline Levels for Selected Airborne Contaminants, Vol. 4. Washington, DC: The National Academies Press.

NRC (National Research Council). 2004b. Research Priorities for Airborne Particulate Matter. IV. Continuing Research Progress. Washington, DC: The National Academies Press.

NRC (National Research Council). 2005. Health Implications of Perchlorate Ingestion. Washington, DC: The National Academies Press.

OEHHA (Office of Environmental Health Hazard Assessment). 2001. Public Health Goal for Benzene in Drinking Water. Office of Environmental Health Hazard Assessment, California Environmental Protection Agency. June 2001 [online]. Available: http://www.oehha.ca.gov/water/phg/pdf/ BenzeneFinPHG.pdf [accessed March 24, 2005].

OEHHA (Office of Environmental Health Hazard Assessment). 2004a. Public Health Goal for Arsenic in Drinking Water. Office of Environmental Health Hazard Assessment, California Environmental Protection Agency. April 2004 [online]. Available: http://www.oehha.ca.gov/water/phg/pdf/ asfinal.pdf [accessed March 24, 2005].

OEHHA (Office of Environmental Health Hazard Assessment). 2004b. Public Health Goal for Perchlorate in Drinking Water. Office of Environmental Health Hazard Assessment, California Environmental Protection Agency. March 2004 [online]. Available: http://www.oehha.ca.gov/water/phg/pdf/ finalperchlorate31204.pdf [accessed March 24, 2005].

OEHHA (Office of Environmental Health Hazard Assessment). 2004c. Recommendation for an Ambient Air Quality Standard for Ozone. Office of Environmental Health Hazard Assessment, California Environmental Protection Agency. June 2004 [online]. Available: http://www.oehha.ca. gov/air/criteria_pollutants/pdf/ozonerec1.pdf [accessed March 29, 2005].

OEHHA/ALA (Office of Environmental Health Hazard Assessment and American Lung Association of California). 2001. Health Effects of Diesel Exhaust. Office of Environmental Health Hazard Assessment, California Environmental Protection Agency [online]. Available: http://www.oehha. ca.gov/public_info/facts/pdf/diesel4-02.pdf [accessed March 24, 2005].

Ott, M.G., B.B. Holder, and H.L. Gordon. 1974. Respiratory cancer and occupational exposure to arsenicals. Arch. Environ. Health. 29(5):250-255.

Pandya, R.J., G. Solomon, A. Kinner, and J.R. Balmes. 2002. Diesel exhaust and asthma: Hypotheses and molecular mechanisms of action. Environ. Health Perspect. 110(Suppl. 1):103-112.

Pavuk, M., A.J. Schecter, F.Z. Akhtar, and J.E. Michalek. 2003. Serum 2,3,7,8-tetrachlorodibenzo-p-dioxin (TCDD) levels and thyroid function in Air Force veterans of the Vietnam War. Ann. Epidemiol. 13(5):335-343.

Peluso, M., A. Munnia, G. Hoek, M. Krzyzanowski, F. Veglia, L. Airoldi, H. Autrup, A. Dunning, S. Garte, P. Hainaut, C. Malaveille, E. Gormally, G. Matullo, K. Overvad, O. Raaschou-Nielsen, F. Clavel-Chapelon, J. Linseisen, H. Boeing, A. Trichopoulou, D. Trichopoulos, A. Kaladidi, D. Palli, V. Krogh, R. Tumino, S. Panico, H.B. Bueno-De-Mesquita, P.H. Peeters, M. Kumle, C.A. Gonzalez, C. Martinez, M. Dorronsoro, A. Barricarte, C. Navarro, J.R. Quiros, G. Berglund, L. Janzon, B. Jarvholm, N.E. Day, T.J. Key, R. Saracci, R. Kaaks, E. Riboli, and P. Vineis. 2005. DNA adducts and lung cancer risk: A prospective study. Cancer Res. 65(17): 8042-8048.

Perera, F., D. Tang, R. Whyatt, S.A. Lederman, and W. Jedrychowski. 2005. DNA damage from polycyclic aromatic hydrocarbons measured by benzo[a]pyrene-DNA adducts in mothers and newborns from Northern Manhattan, the World Trade Center Area, Poland, and China. Cancer Epidemiol. Biomarkers Prev. 14(3):709-714.

Pope, C.A., III, M.J. Thun, M.M. Namboodiri, D.W. Dockery, J.S. Evans, F.E. Speizer, and C.W. Heath, Jr. 1995. Particulate air pollution as a predictor of mortality in a prospective study of U.S. adults. Am. J. Respir. Crit. Care Med. 151(3 Pt 1):669-674.

Pope, C.A., III, R.T. Burnett, M.J. Thun, E.E. Calle, D. Krewski, K. Ito, and G.D. Thurston. 2002. Lung cancer, cardiopulmonary mortality, and long-term exposure to fine particulate air pollution. JAMA 287(9):1132-1141.

Pope, C.A., III, R.T. Burnett, G.D. Thurston, M.J. Thun, E.E. Calle, D. Krewski, and J.J. Godleski. 2003. Cardiovascular mortality and long-term exposure to particulate air pollution: Epidemiological evidence of general pathophysiological pathways of disease. Circulation 109(1):71-77.

Rencher, A.C., M.W. Carter, and D.W. McKee. 1977. A retrospective epidemiological study of mortality at a large western copper smelter. J. Occup. Med. 19(11):754-758.

Rinsky, R.A., R.J. Young, and A.B. Smith. 1981. Leukemia in benzene workers. Am. J. Ind. Med. 2(3):217-245.

Rinsky, R.A., A.B. Smith, R. Hornung, T.G. Filloon, R.J. Young, A.H. Okun, and P.J. Landrigan. 1987. Benzene and leukemia: An epidemiologic risk assessment. N. Engl. J. Med. 316(17):1044-1050.

Roth, F. 1958. Bronchial cancer of arsenic-poisoned vintagers [in German]. Virchows. Arch. 331(2):119-137.

Samet, J., and D. Krewski. 2005. Health effects associated with exposure to ambient air pollution. Pp. 55-76 in Strategies for Clean Air and Health, L. Craig, D. Krewski, J. Shortreed, and J. Samet, eds. Institute for Risk Research, University of Waterloo, Waterloo, Ontario [online]. Available: http://www.irr-neram.ca/rome/Proceedings/Samet.pdf [accessed Oct. 5, 2005].

Samet, J.M., F. Dominici, F.C. Curriero, I. Coursac, and S.L. Zeger. 2000. Fine particulate air pollution and mortality in 20 U.S. cities, 1987-1994. N. Engl. J. Med. 343(24):1742-1749.

Schroeder, J.C., K. Conway, Y. Li, K. Mistry, D.A. Bell, and J.A. Taylor. 2003. p53 mutations in bladder cancer: Evidence for exogenous versus endogenous risk factors. Cancer Res. 63(21):7530-7538.

Schurdak, M.E., and K. Randerath. 1989. Effects of route of administration on tissue distribution of DNA adducts in mice: Comparison of 7H-dibenzo(c,g)carbazole, benzo(a)pyrene, and 2-acetylaminofluorene. Cancer Res. 49(10):2633-2638.

Shao, L, S.L. Lerner, J. Bondaruk, B.A. Czerniak, X. Zeng, H.B. Grossman, M.R. Spitz, and X. Wu. 2004. Specific chromosome aberrations in peripheral blood lymphocytes are associated with risk of bladder cancer. Genes Chromosomes Cancer 41(4):379-389.

Singh, R.S. 2003. Darwin to DNA, molecules to morphology: The end of classical population genetics and the road ahead. Genome 46(6):938-942.

Siwinska, E., D. Mielzynska, and L. Kapka. 2004. Association between urinary 1-hydroxypyrene and genotoxic effects in coke oven workers. Occup. Environ. Med. 61(3):e10.

Thun, M.J., T.M. Schnorr, A.B. Smith, and W.E. Halperin. 1985. Mortality among a cohort of U.S. cadmium production workers: An update. J. Natl. Cancer Inst. 74(2):325-333.

Tokudome, S., and M. Kuratsune. 1976. A cohort study on mortality from cancer and other causes among workers at a metal refinery. Int. J. Cancer 17(3):310-317.

Turteltaub, K.W., C.E. Frantz, M.R. Creek, J.S. Vogel, N. Shen, and E. Fultz. 1993. DNA adducts in model systems and humans. J. Cell Biochem. Suppl. 17F:138-148.

Walton, G. 1951. Survey of literature relating to infant methemoglobinemia due to nitrate-contaminated water. Am. J. Public Health 41(8:2):986-996.

Watson, W.A., T.L. Litovitz, W. Klein-Schwartz, G.C. Rodgers, Jr., J. Youniss, N. Reid, W.G. Rouse, R.S. Rembert, and D. Borys. 2004. 2003 Annual report of the American Association of Poison Control Centers Toxic Exposure Surveillance System. Am. J. Emerg. Med. 22(5):335-404.

WHO (World Health Organization). 2001. Biomarkers in Risk Assessment: Validity and Validation. Environmental Health Criteria 222. Geneva: World Health Organization.

Whyatt, R.M., V. Rauh, D.B. Barr, D.E. Camann, H.F. Andrews, R. Garfinkel, L.A. Hoepner, D. Diaz, J. Dietrich, A. Reyes, D. Tang, P.L. Kinney, and F.P. Perera. 2004. Prenatal insecticide exposures and birth weight and length among an urban minority cohort. Environ. Health Perspect. 112 (10):1125-1132.

Wiencke, J.K., K.T. Kelsey, A. Varkonyi, K. Semey, J.C. Wain, E. Mark, and D.C. Christiani. 1995. Correlation of DNA adducts in blood mononuclear cells with tobacco carcinogen-induced damage in human lung. Cancer Res. 55(21):4910-4914.

Wu-Williams, A.H., L. Zeise, and D. Thomas. 1992. Risk assessment for aflatoxin B1: A modeling approach. Risk Anal. 12(4):559-567.

4

Strategies for Toxicity Testing

Toxicity-testing strategies designed to generate information on potential hazards or risks posed by environmental agents have evolved in response to legislative mandates, scientific developments, and public concerns. Accordingly, testing strategies have been developed or mandated to evaluate pesticides, food additives, high-production-volume (HPV) industrial chemicals in the United States and the European Union, and risks associated with endocrine-disrupting chemicals, developmental toxicants, and carcinogens. The widespread inclusion of genetic-toxicity tests in testing schemes followed scientific advancements that led to the understanding and general recognition that chemicals could cause mutations and mutations could cause cancer. Concerns over other specific effects have also led to modification and refinement of testing strategies so that chemicals posing those hazards would be identified. For example, the thalidomide and diethylstilbestrol disasters emphasized the need for testing strategies to assess chemically mediated effects on reproduction and development. Furthermore, concerns over the contribution of environmental agents to the development of neurobehavioral disorders, neurodegenerative diseases, and respiratory disorders have highlighted the need for testing strategies to address those possible effects (Peters et al. 1999; Gauderman et al. 2000; Schettler 2001). In addition to identifying hazards, toxicity-testing strategies can be designed to provide a basis for dose-response assessment (for establishing exposures expected to pose no risk or for estimating the relationship between risk and exposure). The ideal testing strategies would provide a systematic approach to gathering the data necessary for hazard identification and dose-response assessment thoroughly, rapidly, at low cost, and with few animals.

In this chapter, the committee characterizes testing strategies used by federal and international agencies to gather the data used to identify and evaluate human health hazards and risks. Several examples of testing strategies that are currently used or have been proposed are presented. The testing strategies identified are not meant to be exhaustive but to illustrate the array of toxicity tests that may be required under different circumstances. Furthermore, presentation of the examples in this chapter is not meant to be a committee endorsement of any given strategy. The chapter concludes with committee observations on the current or proposed strategies.

TYPES OF TOXICITY-TESTING STRATEGIES

In practice, testing strategies vary considerably, although they can often be described by three basic testing approaches: battery, tiered, and tailored. A battery is a specific set of toxicity tests applied to all chemicals in a group. Testing batteries are sometimes intended to provide the minimal dataset necessary for risk-informed regulation or risk management.

In tiered testing, the results of a specific set of toxicity tests and risk-management needs are used to guide decisions about the nature and extent of further testing. Often, a substance is first assigned to categories (for example, based on structure or exposure) that guide testing sequences. The chemical then moves through a series of tests sequentially with the data from each test informing the next step in the process.

In tailored testing, information on exposure circumstances, suspected adverse effects, and knowledge of mechanism of action is used to determine the scope of tests to be conducted on a given chemical or class of chemicals. The strategies are thus tailored to the nature of the substance under consideration, its likely use or the likely exposure to it, and the extent of the information available and information needed. Tailored testing strategies may start with a flexible test battery and evolve to different tiers or types of testing in an iterative manner based on scientific judgment.

Characterizing an overall testing strategy as a battery, tiered, or tailored approach is often not possible, because testing strategies are typically combinations of these three basic elements. The examples that follow illustrate that point and demonstrate that testing strategies differ based on the concerns that they were meant to address. The examples address three applications: testing for registration of pesticides and food additives, screening of large numbers of chemicals to develop basic hazard and risk information, and screening of chemicals for specific effects of concern, such as endocrine disruption.

TOXICITY-TESTING STRATEGIES
FOR PESTICIDES AND FOOD ADDITIVES

To protect the food supply, a pesticide cannot be sold or distributed in the United States without being licensed by the Environmental Protection Agency (EPA), and food additives must be formally approved for use by the Food and Drug Administration (FDA). The exceptions are food additives that had already been sanctioned as safe when the 1958 food-additive amendments were adopted or that were generally recognized as safe by FDA. To provide a basis for evaluating the safety of pesticides and food additives, FDA and EPA require a series of tests from applicants and petitioners, as discussed in the following sections.

Federal Insecticide, Fungicide, and
Rodenticide Act Testing Program

The EPA Office of Pesticide Programs regulates the use of pesticides under the authority of the Federal Insecticide, Fungicide, and Rodenticide Act (FIFRA). Through FIFRA, EPA mandates a battery of toxicity tests of conventional chemical pesticides to assist in determining the precautionary language that is required on the label, the type of personal protective equipment that is required for appliers, and the types of uses and use rates to allow. The tests (Table 4-1) are grouped in five main categories: acute tests, subchronic tests, chronic tests, mutagenicity tests, and special tests. Most of the tests are required for pesticides that can end up as food residues or potentially have widespread exposure of the general population, such as those which have residential use. For other pesticides, only acute and mutagenicity testing may be required. Additional studies—for example, dermal penetration, 21-day dermal, subchronic dermal, subchronic inhalation, acute and subchronic neurotoxicity, acute and subchronic delayed neurotoxicity, and developmental neurotoxicity—may be triggered by some special characteristic of a pesticide (such as its chemical class), by potential use and exposure patterns (such as residential uses), or by the results of routinely required studies.

TABLE 4-1 Battery of Tests Required by EPA for New Pesticide Chemicals

Tests	Food Uses	Nonfood Uses
Acute tests		
Acute oral toxicity—rat	R	R
Acute dermal toxicity	R	R
Acute inhalation toxicity—rat	R	R
Primary eye irritation—rabbit	R	R
Primary dermal irritation	R	R
Dermal sensitization	R	R
Delayed neurotoxicity—hen	R	R
Subchronic testing		
90-day feeding studies—rodent and nonrodent	R	C
21-day dermal toxicity	C	C
90-day dermal toxicity	C	C
90-day inhalation—rat	C	C
90-day neurotoxicity—hen or mammal	C	C
Chronic tests		
Chronic feeding of two species—rodent and nonrodent	R	C
Oncogenicity study of two species—rat and mouse preferred	R	C
Teratogenicity in two species	R	C
Reproduction—two-generation	R	C
Mutagenicity tests		
Gene mutation	R	R
Structural chromosomal aberration	R	R
Other genotoxic effects	R	R
Special tests		
General metabolism	R	C
Dermal penetration	C	C
Domestic animal safety	C	C

Note: R = required data; C = conditionally required data on the basis of special pesticide characteristics, potential use and exposure patterns, or results of routinely required studies.
Source: Adapted from 40 CFR 158.340.

EPA has the authority to impose data requirements on pesticides beyond what is required routinely if it determines that more data are needed to characterize the hazard potential of a particular pesticide, including potential hazards to infants and children.

Microbial Pest-Control Agents Testing Program

Microbial pest-control agents (MPCAs) are natural and strain-improved bacteria, algae, fungi, viruses, and protozoa that act as biologic pesticides (40CFR152.20[2002]). MPCAs typically have unique or nontoxic modes of action and are often naturally occurring. They are most appropriately characterized for health and environmental safety with testing schemes that take their unique characteristics into account. Unlike chemical pesticides, MPCAs may survive and reproduce in the environment and may infect or cause disease in other living organisms. Consequently, basic testing protocols are designed specifically to detect any of those characteristics. Protocols for further testing emphasize exposure or environmental expression in addition to infectivity and pathogenicity.

Toxicity-testing requirements, as described by EPA (2004), are set forth in two tiers (see Table 4-2). Tier I consists of a battery of short-term tests designed to evaluate toxicity, infectivity, and pathogenicity. Tier II is designed to evaluate the particular situation when either toxicity or infectivity but not pathogenicity is observed in tier I (EPA 1996).

FDA Testing Strategies for Food Additives

FDA provides guidance to industry and the public concerning the procedures and methods for assessing the safety of direct and indirect food and color additives. FDA published its guidelines, *Toxicological Principles for the Safety Assessment of Food Ingredients*, also known as the Redbook, in 1982 and revised them in 1993, 2000, and 2004. The

TABLE 4-2 Toxicity Tests for Microbial Pest-Control Agents

Tier I	Tier II
Mammalian toxicity, pathogenicity, infectivity	Acute toxicology
Acute oral toxicity, pathogenicity	Subchronic toxicity, pathogenicity
Acute dermal toxicity, pathology	Reproductive and fertility effects
Acute pulmonary toxicity, pathogenicity	
Acute injection toxicity, pathogenicity	
Hypersensitivity incidents	
Cell culture	

Source: EPA 2004.

Redbook contains general guidelines for toxicity studies, including such issues as good laboratory practices, test-animal housing and maintenance, species selection, age, diet, observations, clinical tests, and histopathologic examinations. The Redbook also takes a prescriptive approach, defining in detail how to test various food ingredients, how the agency will review the data, and how decisions will be made. It stipulates specific study designs, what data to collect and report, and how pathologic and statistical analyses should be used in data interpretation. Traditional toxicologic methods are emphasized. FDA states that sponsors may use alternatives to methods that are contained in the Redbook as long as they satisfy applicable regulations and statutes.

The process of evaluating a direct food additive begins with assignment of the additive to one of three concern levels on the basis of chemical structure and expected concentrations of the chemical in the diet. Specifically, the additive is first assigned to structure category A, B, or C (FDA 1993). Structure category A includes compounds associated with low toxic potential or identified as normal cellular constituents, such as alkanes, complex carbohydrates, and fatty acids. Structure category B includes compounds associated with noncancer adverse effects in animals or humans, such as certain amino acids, carboxylic anhydrides, peptides, and proteins. Structure category C includes compounds associated with mutagenicity or carcinogenicity, such as benzofurans, epoxides, and phenols. Potential exposure based on dietary concentration determines the final concern level. This process is illustrated in Table 4-3. Indirect food additives—chemicals that become part of food in trace amounts because of packaging, storage, or other handling—are categorized for testing only according to dietary concentration.

TABLE 4-3 Concern Levels for Direct Food Additives[a]

	Degree of Concern	
Higher ←		→ Lower
Concern Level III	Concern Level II	Concern Level I
Structure C 0.25 ppm	Structure C 0.0125 ppm	Structure C <0.0125 ppm
Structure B 0.5 ppm	Structure B 0.025 ppm	Structure B <0.025 ppm
Structure A 1.0 ppm	Structure A 0.05 ppm	Structure A <0.05

[a]All concentrations listed in the table are estimated concentrations in the total diet.
Source: Adapted from FDA 1997, 2003.

As shown in Table 4-4, specific tests are required on the basis of concern level for direct food additives or dietary concentration for indirect food additives. For example, a short-term feeding study in rodents and short-term tests for carcinogenic potential would be required for direct food additives with the lowest concern (concern level I). For an indirect food additive expected to occur at less than 0.05 ppm in the diet, only an acute oral toxicity study in rodents would be required. The tests increase in complexity and duration as the level of concern or dietary concentration increases. As Table 4-4 indicates, some tests are contingent on other test findings, but there is no formal guidance on proceeding

TABLE 4-4 Testing Required for Direct and Indirect Food Additives

Direct Food Additives		Toxicity Tests
Concern Level[a]	Testing Required	A. Acute oral study—rodent
I	B, K	B. Short-term feeding study (at least 28 days)—rodent
II	A***, D, E, I, J*, K	C. Subchronic feeding study (90 days)—rodent with in utero exposure
III	A***, D***, F, G, H, I, J*, K, L**	D. Subchronic feeding study (90 days)—rodent
Indirect Food Additives		E. Subchronic feeding study (90 days)—nonrodent
Dietary Concentration	Testing Required	F. Lifetime feeding study (about 2 years)—rodent with in utero exposure for carcinogenesis and chronic toxicity
<0.05 ppm	A	
>0.05 ppm	A***, C, I*, K**, E, J*	G. Lifetime feeding study (about 2 years)—rodent for carcinogenesis
>1.0 ppm	A***, D***, F, G, H, I, J*, K**, L**	H. Short-term feeding study (at least 1 year)—nonrodent
		I. Multigeneration reproduction feeding study (at least two generations) with teratology phase—rodent
* If indicated by available data or information. ** Suggested. *** If needed as preliminary to further study.		J. Teratology study
		K. Short-term tests for carcinogenic potential
		L. Metabolism studies
		The Redbook contains references to current guides on these tests.

[a]The concern level is determined by chemical structure and dietary concentration (see Table 4-3).
Source: Adapted from FDA 1997.

from one test to another on the basis of negative or positive results. FDA may request studies that go beyond its guidance and does mention epidemiologic studies for assessing safety, but it is vague as to when they could best be used.

FDA's approach to the hazard assessment of drug candidates is somewhat different from its approach for food additives. Although FDA expects a package of toxicity studies similar to those required for food additives, it is far more willing to modify pharmacologic and toxicologic studies so that they answer questions peculiar to the proposed therapeutic entity. Thus, safety assessments of drugs are less prescriptive than those of food ingredients. They lack detail on study designs but focus on the principles that need to be addressed.

TOXICITY-TESTING STRATEGIES FOR
SCREENING OF INDUSTRIAL CHEMICALS

In addition to the vast number of chemicals already in commerce, many chemicals are introduced each year. The following describes testing strategies used in the United States and Europe to screen and provide the basis for toxicity assessment of new and existing industrial chemicals.

The Toxic Substances Control Act and the
High-Production-Volume Chemical Testing Program

The Toxic Substances Control Act (TSCA) was passed in 1976 and gave EPA the authority to collect information and issue regulations on new and existing industrial chemical substances. When EPA began evaluating chemicals under TSCA in 1979, about 62,000 chemicals were in commerce (GAO 2005). Today, 82,000 chemicals are in commerce, and about 700 chemicals are introduced each year (GAO 2005).

TSCA does not prescribe or detail a testing strategy to evaluate the large volume of existing and new chemicals. Companies that manufacture or process new chemicals for commercial purposes must submit to EPA a premanufacturing notice (PMN), which includes information on chemical structure, production process, expected production volume, intended uses, possible exposure and release levels, disposal procedures, and other data "concerning the chemical's environmental or health effects known to or reasonably ascertainable by the chemical company"

(GAO 2005). To register a new chemical, companies are not required to conduct any specific toxicity tests, and EPA estimates that only about 15% of PMNs contain health or safety data (GAO 2005). EPA typically uses models to predict a chemical's toxicity on the basis of its structure. A more detailed review is conducted only on about 20% of the chemicals (GAO 2005). EPA does have the authority to require manufacturers, importers, and processors of chemical substances to submit new data on existing chemicals, and it can issue a test rule that indicates specific tests to be conducted. However, the process is burdensome, and "EPA has used its authority to require testing for fewer than 200 of the 62,000 chemicals in commerce when EPA began reviewing chemicals under TSCA in 1979" (GAO 2005).

To address the lack of data on existing industrial chemicals, EPA implemented a voluntary program in negotiation with the American Chemistry Council, the American Petroleum Institute, and Environmental Defense known as the HPV chemical testing program. The program's purpose is to ensure that basic toxicity data are available on all organic, nonpolymeric chemicals produced or used in the United States in excess of 1 million pounds per year. EPA focused attention on HPV chemicals because it considered such chemicals to have a higher potential for environmental and workplace exposure than low-production-volume chemicals (65 Fed. Reg. 81661[2000]). The HPV testing program is intended to support the development of screening-level hazard and risk characterizations with a battery of tests for basic toxicity testing end points (see Box 4-1). EPA is implementing the program to be consistent with the HPV program developed by the Organisation for Economic Co-operation and Development (OECD) in that it includes the

---

**BOX 4-1 Human Health Data to Be Obtained
with Basic SIDS Testing Battery**

- Acute toxicity
- Repeated-dose toxicity
- Genetic toxicity in vitro
  —Point mutation
  —Chromosomal aberration
- Genetic toxicity in vivo (provisional)
- Reproductive toxicity
- Developmental toxicity and teratogenicity

same end points and extent of testing as the OECD screening information dataset (SIDS) (65 Fed. Reg. 81686 [2000]).

The international community has agreed that SIDS is the minimal dataset required to screen HPV chemicals for toxicity (OECD 2004a). In the United States, if SIDS toxicity data are not available or are incomplete for a particular chemical, a test plan is proposed and reviewed by EPA and other organizations. Test plans can be designed for categories of chemicals that have common physiochemical characteristics, common functional groups, and common toxic properties. Specific tests are conducted according to EPA or comparable OECD test guidelines (65 Fed. Reg. 81686 [2000]). Tests are sponsored by private organizations, and the results are made publicly available. EPA (65 Fed. Reg. 81686 [2000]) notes that the results of the SIDS testing will support preliminary risk assessment, anticipating that for some chemicals of lower concern the results will be adequate to evaluate the hazards and risks posed. The results may indicate the need for further testing of other chemicals. Conceptually, the SIDS battery is an initial battery in an overall tiered approach to testing for hazard and risk assessment.

Canada has a program similar to the U.S. and OECD HPV testing programs (GAO 2005).

European Union Testing Strategies

Toxicity testing of environmental agents to inform human health risk assessments occurs in the European Union (EU) under Commission Directives 93/67/EEC and 98/8/EC and Commission Regulation 1488/94, which direct the risk assessment of new substances, existing substances, and biocidal products. A human health risk assessment must be carried out for all existing substances (substances marketed before September 18, 1981) and for new substances that are identified because of their toxic or physiochemical properties and possible human exposure via workplace, product consumption, or indirect environmental exposure.

The EU issued a technical guidance document (TGD) on risk assessment (EC 2003) for use by authorities to help in carrying out toxicity testing. EC (2003) describes a process whereby chemicals in the EU are tested according to their annual manufactured quantity; thus, it is a tonnage-driven testing program. That strategy is based on the assumption that adequate risk assessment of chemicals with low general exposure might not require as much toxicity-testing data as risk assessment of substances with higher exposure. The scheme is intended to assess the risk posed to humans by individual substances; additive and synergistic effects caused by the combined action of several substances are not considered.

The TGD details different sets of toxicity tests that are required on the basis of the quantity of substance produced per manufacturer per year. All substances that are produced at more than 10 kg/manufacturer per year must be tested (see Table 4-5). Tests of new substances are to be carried out in accordance with EU test guidelines (Annex V to Directive 67/548) or, if EU guidelines are not available, OECD guidelines.

STRATEGIES FOR SCREENING CHEMICALS FOR EFFECTS ON SPECIFIC SYSTEMS AND END POINTS

Because of the potential widespread impact of introducing into the environment anthropogenic materials that disrupt the endocrine system, EPA is developing strategies to screen chemicals for endocrine-disrupting activity, and OECD has developed a general framework for such evaluations. The National Research Council (NRC) undertook a study to clarify how environmental agents may be affecting human development and made recommendations for improvements to qualitative and quantitative risk assessment. The EPA and OECD endocrine-disruptor screening approaches and work of the NRC are provided as examples of testing approaches aimed at addressing specific systems and end points.

Environmental Protection Agency
Endocrine-Disruptor Testing Strategy

In 1996, EPA formed the Endocrine Disruptor Screening and Testing Advisory Committee (EDSTAC) to advise on the development of a program for screening and testing chemicals for endocrine-disrupting activity.[1] After considering that committee's advice, EPA (63 Fed. Reg.

---

[1]In 1995, EPA, the U.S. Department of the Interior, the U.S. Centers for Disease Control and Prevention, and the U.S. Congress asked the National Research Council (NRC) to conduct an independent evaluation of the potential adverse effects of environmental exposure to endocrine disruptors. The primary focus of the report (NRC 1999) was evaluation of potential reproductive, developmental, neurologic, immunologic, and carcinogenic effects of suspected endocrine disruptors. The report did not provide a specific framework analogous to the work of the EDSTAC.

TABLE 4-5 European Union Testing Strategy

10–100 kg/manufacturer per year	1–10 tons/manufacturer per year	10–1,000 tons/manufacturer per year (50–5,000 tons cumulative)
Acute toxicity—oral or inhalation	*Acute toxicity*—oral route and a second depending on likely route of exposure	Level 1 Base Testing Set
100 kg–1 ton/manufacturer per year	*Skin irritation*	*Acute toxicity*—oral route and a second depending on likely route of exposure
Acute toxicity—oral route and a second depending on likely route of exposure	*Eye irritation*	*Skin irritation*
Skin irritation	*Skin sensitization*	*Eye irritation*
Eye irritation	*Repeated-dose toxicity—28 day*	*Skin sensitization*
Skin sensitization	*Mutagenicity*—bacteriologic and nonbacteriologic tests	*Repeated-dose toxicity—28 day*
Mutagenicity—bacteriologic test	*Reproductive-toxicity screen*	*Mutagenicity*—bacteriologic and nonbacteriologic tests
Reproductive-toxicity screen	*Pharmacokinetic assessment*—based on data derived from above tests	*Reproductive-toxicity screen*
Pharmacokinetic assessment—based on data derived from above tests		*Pharmacokinetic assessment*—based on data derived from above tests
		Fertility study—in one species for one generation
		May be required on basis of level 1 testing results
		Fertility—second-generation study
		Teratogenicity—in second species
		Prenatal-development toxicity
		Subchronic/chronic toxicity
		Mutagenicity—additional testing

(Continued)

TABLE 4-5 *Continued*

**>1,000+ tons/manufacturer per year
(>5,000 tons cumulative)**

Level 2 Base Testing Set

Acute toxicity—oral route and a second
depending on likely route of exposure

Skin irritation

Eye irritation

Skin sensitization

Repeated-dose toxicity—28 days

Mutagenicity—bacteriologic and
nonbacteriologic tests

Reproductive-toxicity screen

Fertility study—in one species for one
generation

Chronic toxicity

Carcinogenicity

Pharmacokinetic assessment—based on
data derived from base testing set plus
additional studies

May be required on basis of level 1 testing
results

Fertility—multigeneration study

Prenatal-development toxicity

Acute toxicity—in second species

Repeated-dose toxicity—28 days—in
second species

42852 [1998]; 63 Fed. Reg. 71542 [1998]) developed its Endocrine Disruptor Screening Program (EDSP), which recommended screening of chemicals that have the potential to disrupt androgen, estrogen, and thyroid hormone systems. Both potential human and ecologic effects were to be addressed, and nonpesticide chemicals, contaminants, and mixtures were to be included in addition to pesticides. EPA developed a tiered approach for the EDSP. Its core elements are priority-setting—tier 1 screening to identify agents with the potential to alter the estrogen, androgen, or thyroid hormone systems and tier 2 testing to determine whether agents identified in tier 1 cause developmental toxicity through any of the three systems and, if so, their dose-response relationships.

Figure 4-1 shows an overview of the EDSP. The EDSTAC estimated that 87,000 chemical agents in use were candidates for screening for endocrine-disrupting activity. That would overwhelm the resources available for screening, so EPA proposed a "compartment-based approach" in which exposure and effects would be used for setting priorities for initial screening. The initial sort was intended to direct chemicals to one of four categories:

- *Category 1.* A "hold" category comprising polymers with molecular weight greater than 1,000 daltons considered unlikely to cross

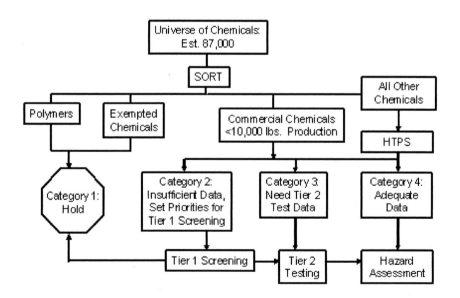

FIGURE 4-1 Endocrine-disruptor screening program overview. Source: EPA 2000.

cell membranes—estimated to amount to 25,000 of the 87,000 chemical agents.

- *Category 2.* Chemicals with insufficient data to undergo tier I high-throughput screening that is designed to detect potential hormonal or biologic activity (binding to estrogen, androgen, or thyroid hormone receptors to elicit a biologic effect, such as transcription activation).
- *Category 3.* Chemicals with existing data sufficient to bypass tier 1 screening and proceed directly to tier 2 testing.
- *Category 4.* Chemicals with existing data sufficient to bypass both tier 1 and tier 2 testing for direct hazard assessment.

The EDSTAC also recommended that a scheme be developed for setting testing priorities among common mixtures. The EDSTAC recognized the practical impossibility of screening all possible chemical combinations but noted that some widely used mixtures, such as pesticide formulations and gasoline formulations, are potential candidates for testing.

Table 4-6 shows the tests included in each tier. The tier 1 screening battery combines in vitro and in vivo assays meant to detect chemicals that affect hormone systems regardless of mode of action; it is intended to minimize false negatives and false positives, and it includes several

TABLE 4-6 Endocrine-Disruptor Screening Tests[a]

Tier 1 Screening Tests
In vitro tests
Estrogen-receptor binding and transcriptional activation
Androgen-receptor binding and transcriptional activation
Steroidogenesis assay using minced testis
In vivo tests
Rodent 3-day uterotrophic assay—subcutaneous administration
Rodent 20-day pubertal female assay with thyroid end points
Rodent 5- to 7-day Hershberger
Frog metamorphosis
Fish gonadal recrudescence
Tier 2 Screening Tests
Mammalian reproduction—two-generation reproductive-toxicity bioassay in rat
Avian reproduction—two-generation test
Fish reproduction—fish life-cycle test
Invertebrate reproduction in mysids or daphnia
Amphibian development and reproduction

[a]As of September 2005, no tier 1 or tier II tests have been validated (EPA 2005).

taxonomic groups and an array of end points to inform a weight-of-evidence evaluation. The in vivo assays are designed as an integral battery that complements the in vitro assays; they cover a wider array of mechanisms of action and incorporate pharmacokinetic determinants of mechanism of action.

Tier 2 testing is designed to determine the likelihood, nature, and dose-response relationship of the disruption of hormone systems in humans, fish, and wildlife. The tests include a wide range of doses administered through a relevant route of exposure throughout critical life stages and processes. The design of the tests also takes into consideration the possibility that the effects of chemical agents may be latent and thus not observed until later in life; this would be analogous to the effects of diethylstilbestrol. Thus, the tests were designed to encompass two generations and to allow determination of effects on fertility, mating, fetal development, neonatal growth and development, and transition from juvenile to sexual maturity.

Organisation for Economic Co-operation and Development Endocrine-Disruptor Testing Strategy

OECD has also developed a conceptual framework for testing endocrine disruptors. OECD emphasizes that the framework represents a toolbox rather than a testing scheme, with levels of tests and assessments that correspond to different levels of biologic complexity (see Table 4-7). That framework considers all testing data, including in vitro data, structure-activity relationships, and data that may become available from new technologies, such as genomics and proteomics. The lower levels of the testing framework are consistent with the hazard-identification phase of the UN Globally Harmonised System, a system for classifying and communicating chemical hazards (UN 2003). The testing and data development in the higher levels support consideration of dose-response relationships. The OECD endocrine-disruptor testing framework indicates that a chemical can enter at any step on the basis of available data or data requirements and leave the testing framework when available data are sufficient for an assessment.

Developmental-Toxicity Testing

Given the potentially devastating effects of human developmental defects, the NRC convened the Committee on Developmental Toxicity to

TABLE 4-7 OECD Conceptual Framework for Testing Endocrine Disruptors

Level 1 Sorting and prioritization based upon existing information	– Physical chemical properties, e.g., MW, reactivity, volatility, biodegradability – Human and environmental exposure, e.g., production volume, release, use patterns – Hazard, e.g., available toxicological data
Level 2 In vitro assays providing mechanistic data	– ER, AR, TR receptor binding affinity – Transcriptional activation – Aromatase and steroidogenesis in vitro – Aryl hydrocarbon receptor recognition/binding – QSARs – High-throughput prescreens – Thyroid function – Fish hepatocyte VTG assay – Others (as appropriate)
Level 3 In vivo assays providing data about single endocrine mechanisms	– Uterotrophic assay (estrogenic related) – Hershberger assay (androgenic related) – Non-receptor mediated hormone function – Others (e.g., thyroid) – Fish VTG (vitellogenin) assay (estrogenic related)
Level 4 In vivo assays providing data about multiple endocrine mechanisms	– Enhanced OECD 407 (end points based on endocrine mechanisms) – Male and female pubertal assay – Intact male assay – Fish gonadal histopathology assay – Frog metamorphosis assay
Level 5 In vivo assays providing adverse effects data from endocrine and other mechanisms	– 1-generation assay (TG 415 enhanced)[a] – 2-generation assay (TG 416 enhanced)[a] – Reproductive screening test (TG 421 enhanced)[a] – Combined 28 day/reproduction screening test (TG 422 enhanced)[a] – Partial and full life cycle assays in fish, birds, amphibians and invertebrates (developmental and reproduction) [a]Potential enhancement will be considered by Validation Management Group covering mammalian methods

Note 1: Entering and exiting at all levels are possible and depend on the nature of existing information needs for hazard and risk-assessment purposes.
Note 2: In level 5, ecotoxicology should include end points that indicate mechanisms of adverse effects and potential population damage.

Note 3: When a multimodal model covers several of the single-end-point assays, that model would replace those single-end-point assays.

Note 4: The assessment of each chemical should be case by case, taking into account all available information and bearing in mind the function of the framework levels.

Note 5: The framework should not be considered as all-inclusive at present. At levels 3, 4, and 5, it includes assays that are either available or for which validation is under way. The latter are provisionally included. Once developed and validated, they will be formally added to the framework.

Note 6: Level 5 should not be considered as including only definitive tests. Tests included at that level are considered to contribute to general hazard and risk assessment.

Abbreviations: AR, androgen-related; ER, estrogen-related; MW, molecular weight; OECD, Organisation for Economic Co-operation and Development; QSAR, quantitative structure-activity relationship; TG, test guideline; TR, thyroid-hormone-related; VTG, vitellogenin.

Source: OECD 2004b.

evaluate the impact of environmental agents on human development. The committee proposed a multidisciplinary, multilevel, interactive approach to developmental-toxicity testing (NRC 2000, Chapter 8). The basic premise of the approach is that understanding the mechanistic basis of extrapolation between test animals (or in vitro assays) and humans will give risk assessment greater validity. The approach is different from a tiered-testing approach because testing may be initiated at any level and there is no unidirectional triggering of higher-level testing by results obtained at a lower level.

Figure 4-2 illustrates the possible testing scheme proposed by that committee. Testing in model systems is described as taking place on four levels. Level 1 includes molecular, biochemical, and cell-based assays that have high throughput and are expected to provide structure-activity information, relative potency of various chemicals tested in the same assay, insight into mixtures, and estimates of potency across chemical classes and assays. The proposed assays are designed to test chemical-induced alterations in conserved signaling and metabolic pathways. The committee envisioned that the application of the high-throughput screens could address 100,000 chemicals within 1 year. Level 2 proposes the use of nonmammalian animals to evaluate their response during developmental exposure. The fruit fly, nematodes, and zebrafish are mentioned as candidate organisms. Those model organisms could be partially "humanized" in their metabolism (that is, genetically

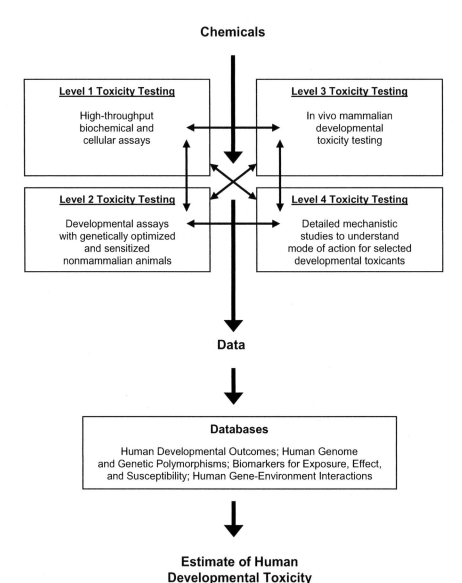

FIGURE 4-2 Developmental-toxicity testing approach.

modified to express human metabolizing proteins) and sensitized to heighten the response of known important developmental pathways. The premise of testing in those models is that for some organs and structures, human and model organism differences are not great. It was estimated that 1,000-10,000 of those assays could be conducted each year. Level 3 involves testing in mammals, predominantly mice and rats, with the assumption that hundreds of assays could be conducted each year. The information expected to be gained in those animal tests is relative in vivo potency, the activity of chemical mixtures, mechanistic information, and dose-response relationships. Level 4 proposes detailed mechanistic evaluations of prototype members of families of chemicals whose mechanisms of action can be elucidated. Genetically optimized rodents are proposed as the likely test platform for those studies. The committee felt that about 10 such detailed evaluations could be conducted each year.

The model system and animal results are correlated with information gained from the assessment of toxicity, susceptibility, and chemical exposures in human populations. In particular, development of a number of databases is proposed, including a database of human developmental outcomes, a database of the human genome and genomic polymorphisms, a database of human biomarkers, and a database of human gene-environment interactions. The proposal recognized the challenge of creating and managing large databases and of linking and interrelating the mass of information.

Another important concept mentioned is the combining of information obtained from the model system and animal tests with that accumulated through an improved human surveillance program. It is less clear how the large mass of information generated in the proposed multilevel testing scheme is translated in practice into risk assessments. The committee's report is presented at a conceptual level and discusses the potential for new technical approaches without providing a detailed plan for change. The committee's proposal does not specify how the current testing paradigms and risk-assessment approaches could be modified and expanded to incorporate the new science and lead to a new scheme for data generation, animal-to-human extrapolation, and risk assessment. However, the committee's report provides a useful framework for thinking about how different types and levels of testing and assessment could lead to an improved overall approach to chemical-toxicity assessments.

COMMITTEE OBSERVATIONS ON
TOXICITY-TESTING STRATEGIES

In 1984, an NRC committee providing advice to the National Toxicology Program on testing priorities noted that there are far more chemicals in the human environment than can be evaluated for potential toxicity with available resources and methods (NRC 1984). That committee bemoaned the fragmentary information available with which to set priorities. The same holds true today. For the roughly 700 new industrial chemicals introduced into commerce each year, EPA essentially relies on its own structure-activity models to assess potential hazards and on information on use and estimated production volume contained in PMNs to characterize potential exposure. TSCA authorizes EPA to review existing chemicals, but toxicity and exposure information on them is typically so incomplete that it does not support the review process. EPA can require testing if it determines that a chemical meets a specific set of criteria; however, in vitro and whole-animal tests are rarely required. Thus, the basis for establishing priorities and requiring testing for industrial chemicals in the United States has not progressed much over the last 20 years.

The HPV program, which was a response to the lack of data collected under TSCA, uses a simple criterion to determine whether a chemical is to be included in the HPV testing program: nonpolymeric organic chemicals produced in or imported into the United States at a volume of 1 million pounds or more during the 1990 reporting year.[2] Similarly, the EU uses production volume to assign priorities and define testing requirements. Unlike the U.S. HPV program, the EU program requires a minimum of testing of all chemicals produced at volumes of more than 10 kg/manufacturer per year. The extent of testing required increases with production volume; substantially greater testing, including carcinogenicity and pharmacokinetic studies, is required for the highest-production chemicals. For moderate- and high-production-volume chemicals, additional tests can be required on the basis of findings in the initial test battery and other tests. That contrasts with the minimal data-set requested in the U.S. HPV program, although EPA has indicated that it intends to use the results of the HPV program to assess further testing needs. For both the U.S. and EU HPV programs, it is unclear how other

---

[2]There are exclusion criteria. For example, a chemical being handled under the OECD program is excluded from consideration in the U.S. HPV program.

exposure considerations—such as use, environmental persistence, and possible chemical breakdown to more active forms—might modify test selection and extent of testing and whether they have been sufficiently considered. Those other exposure considerations may lead to a better understanding of human exposure potential than production data alone.

Exposure considerations also dictate the degree of testing of direct and indirect food additives and pesticides. Testing is required for virtually all pesticides and all food additives that had not been sanctioned before the 1958 Food Additives Amendments or were not generally recognized as safe. Exposure is qualitatively considered in determining pesticide testing and is based on use. Potential exposures of the general population, especially via food, drive the greatest extent of testing. For indirect and direct food additives, a quantitative assessment of exposure (that is, concentration of the additive in the total diet) provides the basis of test selection. For direct food additives, study selection is modified by information on chemical structure. Regardless of the extent to which the "bright lines" used for food additives provide adequate protection, the overall approach of combining structural alerts with quantitative exposure information is an aspect of test-strategy design to explore for environmental agents.

The testing strategies discussed above generally use traditional, standardized toxicology tests. Results of an initial series of mandated tests can trigger additional testing to provide broader coverage of end points, exposure routes, or life stages or a greater depth of understanding of the effects observed. The tests that are triggered are also standardized toxicology studies.

Information generated by the testing strategies provides a basis for hazard identification (that is, assessment of whether a chemical has the potential to cause specific adverse responses regardless of exposure context). The test results indicate the exposures that produce adverse effects and thus help to characterize the dose-response relationship at least in the range of the experimental data, typically at high doses. The information produced is often sufficient for decision-making. For example, test results may indicate that a food additive or pesticide can be safely used as proposed. Given the doses at which toxicity is seen and the range of possible human exposures, risks may be so low that there is no need to curtail exposure even if one considers possible end points and life stages missed by the testing strategy. In contrast, if the margin between toxic doses and relevant human exposures is not large enough, further research may be needed to refine the dose-response relationship at lower doses, particularly if some stakeholders find that the chemical is worth trying to keep on the market.

Thus, findings of toxicity studies can provoke a series of voluntary studies—often sponsored by industry or regulatory, scientific, or public health agencies—that seek to develop a better qualitative and quantitative understanding of dose-response characteristics and therefore a better understanding of the relevance of the findings for humans. The tests can be nonstandard and tailored to answer specific questions, such as those concerning mechanism of action or species differences in metabolism. The selection of tests and lines of inquiry can be ad hoc without a general framework for inquiry. For example, a general framework has not been adopted for generating and testing hypotheses and alternatives regarding mechanism of toxic action. Much to the frustration of sponsors, results of such testing can be found unacceptable for regulatory applications and not usable for refining risk estimates, although in several notable cases results have been found to be definitive. The ad hoc and unsystematic approaches for generating data for more detailed risk assessment contrast sharply with the mandated strategies for pesticide and food-additive registration and approval.

The toxicity tests and strategies discussed here have evolved primarily as a means of characterizing potential human health hazards and dose-response relationships at least at high doses. Existing strategies only partially recognize the toxicity-testing problems associated with exposure to mixtures and the associated problems in assessing aggregate exposures and cumulative risks. Dioxins, organophosphorus insecticides, and environmental estrogens have been tested as mixtures, but most testing continues to focus on individual chemicals. Different testing approaches generally stem from legislative mandates or from differences in individual agencies or program offices. Different approaches can result in inconsistent testing strategies among agencies or categories of chemicals even if the ultimate regulatory goal is the same. The differences in strategies for testing industrial chemicals by the EU and EPA are an example. The nature and extent of toxicity testing ideally would be guided by the regulatory risk-management decisions to be made and the assessments needed to support them.

NRC and OECD multilevel, multidisciplinary frameworks for evaluating developmental toxicants and endocrine disruptors provide insights into improving the standard regulatory framework for data generation and risk assessment. The general frameworks address key areas of uncertainty in cross-species and high-to-low dose extrapolation and suggest approaches for incorporating high-volume screening of chemicals for those end points. Although considerable resources and commitment would be needed to develop such programs, they hold the promise of shifting to an improved system of testing and decision-making.

REFERENCES

EC (European Commission). 2003. Technical Guidance Document on Risk Assessment in Support of Commission Directive 93/67/EEC on Risk Assessment for New Notified Substances; Commission Regulation (EC) No. 1488/94 on Risk Assessment for Existing Substances, and Directive 98/8/EC of the European Parliament and the Council Concerning the Placing of Biocidal Products on the Market, 2nd Ed. EUR 20418 EN/1. European Chemicals Bureau, Institute for Health and Consumer Protection, Ispra, Italy [online]. Available: http://ecb.jrc.it/index.php? CONTENU=/Technical-Guidance-Document/ [accessed March 21, 2005].

EPA (U.S. Environmental Protection Agency). 1996. Discussion of registration, field testing, and experimental use permits. III Toxicology requirements. P. 3(e) in Microbial Pesticide Test Guidelines, OPPTS 885.0001. Overview for Microbial Pest Control Agents. EPA 712-C-96-280. Office of Prevention, Pesticides, and Toxic Substances, U.S. Environmental Protection Agency, Washington, DC [online]. Available: http://www.epa.gov/oppbppd1/biopesticides/regtools/guidelines/oppts_88 5_0001.htm#(e)%20Discussion%20of%20registration [accessed March 21, 2005].

EPA (U.S. Environmental Protection Agency). 2000. P. 8 in Environmental Protection Agency Endocrine Disruptor Screening Program: Report to Congress [online]. Available: http://www.epa.gov/scipoly/oscpendo/docs/ reporttocongress0800.pdf [accessed March 21, 2005].

EPA (U.S. Environmental Protection Agency). 2004. Microbial Pesticide Test Guidelines OPPTS 885, Group C - Toxicology Test Guidelines. Office of Prevention, Pesticides, and Toxic Substances, U.S. Environmental Protection Agency, Washington, DC [online]. Available: http://www.epa.gov/ oppbppd1/biopesticides/regtools/guidelines/microbial_gdlns.htm [accessed March 21, 2005].

EPA (U.S. Environmental Protection Agency). 2005. Endocrine Disruptor Screening Program, Assay Development and Validation, Assay Status Table. Office of Science Coordination and Policy; Office of Prevention, Pesticides, and Toxic Substances, U.S. Environmental Protection Agency, Washington, DC [online]. Available: http://www.epa.gov/scipoly/ oscpendo/assayvalidation/status.htm [accessed Sept. 20, 2005].

FDA (U.S. Food and Drug Administration). 1993. Draft toxicology principals for safety assessment of direct food additives and color additives used in food. "Redbook II." Center for Food Safety and Applied Nutrition, U.S. Food and Drug Administration, Washington, DC.

FDA (U.S. Food and Drug Administration). 1997. Toxicological Testing of Food Additives. Office of Premarket Approval, Center for Food Safety and Applied Nutrition, U.S. Food and Drug Administration [online]. Available: http://www.cfsan.fda.gov/~dms/opa-tg1.html [accessed March 21, 2005].

FDA (U.S. Food and Drug Administration). 2003. General guidelines for designing and conducting toxicity studies. Section IV. B.1. in Redbook 2000, Toxicological Principles for the Safety Assessment of Food Ingredients, Office of Food Additive Safety, U.S. Food and Drug Administration [online]. Available: http://www.cfsan.fda.gov/~redbook/red-ivb1.html [accessed March 21, 2005].

GAO (U.S. Government Accounting Office). 2005. Chemical Regulation: Options Exist to Improve EPA's Ability to Assess Health Risks and Manage its Chemical Review Program. GAO-05-458. U.S. Government Accounting Office, Washington, DC. June 2005 [online]. Available: http://www. gao.gov/new.items/d05458.pdf [accessed Oct. 24, 2005].

Gauderman, W.J., R. McConnell, F. Gilliland, S. London, D. Thomas, E. Avol, H. Vora, K. Berhane, E.B. Rappaport, F. Lurmann, H.G. Margolis, and J. Peters. 2000. Association between air pollution and lung function growth in southern California children. Am. J. Respir. Crit. Care Med. 162(4 Pt 1):1383-1390.

NRC (National Research Council). 1984. Toxicity Testing: Strategies to Determine Needs and Priorities. Washington, DC: National Academy Press.

NRC (National Research Council). 1999. Hormonally Active Agents in the Environment. Washington, DC: National Academy Press.

NRC (National Research Council). 2000. Scientific Frontiers in Developmental Toxicology and Risk Assessment. Washington, DC: National Academy Press.

OECD (Organisation for Economic Cooperation and Development). 2004a. Manual for Investigation of HPV Chemicals. Organisation for Economic Cooperation and Development [online]. Available: http://www.oecd.org/ document/7/0,2340,en_2649_201185_1947463_1_1_1_1,00.html. [accessed Sept. 20, 2005].

OECD (Organisation for Economic Cooperation and Development). 2004b. Conceptual Framework for the Testing and Assessment of Endocrine Disrupting Chemicals. Prepared by the Secretariat of the Test Guidelines Programme, Based on the Agreement Reached at the 6th Meeting of the EDTA Task Force, 24 and 25 June 2002, Tokyo, Japan [online]. Available: http://www.oecd.org/dataoecd/17/33/23652447.doc [accessed March 21, 2005].

Peters, J.M., E. Avol, W.J. Gauderman, W.S. Linn, W. Navidi, S.J. London, H. Margolis, E. Rappaport, H. Vora, H. Gong, Jr., and D.C. Thomas. 1999.

A study of twelve Southern California communities with differing levels and types of air pollution. II. Effects on pulmonary function. Am. J. Respir. Crit. Care Med. 159(3):768-775.

Schettler T. 2001. Toxic threats to neurologic development of children. Environ. Health Perspect. 109(Suppl 6):813-816.

UN (United Nations). 2003. Globally Harmonized System of Classification and Labeling of Chemicals. United Nations Economic Commission for Europe [online]. Available: http://www.unece.org/trans/danger/publi/ghs/ghs_rev00/00files_e.html [accessed Nov. 17, 2005].

5

Use of Data in Human
Health Risk Assessment

Data from toxicity testing and epidemiologic studies are used in risk assessments and in environmental decision-making to identify potential hazards, to rank environmental problems, to determine the need for further cleanup at a contaminated site, and ultimately to establish environmental standards and exposure guidelines. A conceptual framework for risk assessment was proposed by the National Research Council (NRC 1983) in *Risk Assessment in the Federal Government: Managing the Process.* The framework consists of the following four components:

- *Hazard identification.* What kind of adverse effects might a substance cause? For example, does developmental toxicity, neurotoxicity, or cancer result from exposure?
- *Dose response.* What is the risk of effects at different exposure levels? Is there a level below which no effects are observed?
- *Exposure.* How are humans exposed, and at what levels and frequencies do exposures occur?
- *Risk characterization.* Given human exposure scenarios, what is the probability of adverse effects? How does risk vary across the population? What are the uncertainties in our understanding of the risk or safety?

For risk assessment, data from animal studies and, less often, from epidemiologic investigations are evaluated to identify the types of adverse effects that may occur. Dose-response data from such studies are also analyzed to predict exposures in humans that should be without risk or pose no more than some predetermined level of risk (Lowrence 1976).

The U.S. Environmental Protection Agency (EPA) and other institutions have issued risk-assessment guidelines[1] that outline the array of studies used and how they might be interpreted for risk assessment. The guidelines cover specific end points—such as developmental toxicity (EPA 1991), reproductive toxicity (EPA 1996a), and neurotoxicity (EPA 1995)— and, more broadly, noncancer dose-response assessments (EPA 2004a). Risk-assessment guidelines for carcinogens have also been issued that provide detailed guidance on the procedures for assembling evidence and evaluating modes of action to identify carcinogens and for conducting quantitative dose-response assessments (EPA 1986, 1999, 2005a; IARC 2005a). Noncancer and cancer guidelines are fundamentally different in that noncancer end points are typically evaluated with diverse studies on a wide variety of specific outcomes whereas carcinogenicity is evaluated with a specific set of bioassays that focus on the degree to which a chemical might increase neoplasia in different organs and cell types in the body. EPA also has published a review to explain its approach to risk assessment (EPA 2004b).

This chapter first outlines risk-assessment guideline documents for neurotoxicity, developmental toxicity, and reproductive toxicity. It then discusses generic guidelines for noncancer dose-response assessments, including the requirements resulting from the Food Quality Protection Act (FQPA). Cancer risk-assessment guidelines are discussed next. Throughout, the chapter notes the use and limitations of toxicologic and epidemiologic data typically available for drawing conclusions about hazards, dose-response relationships, and risk based on the guidelines. The chapter focuses on current institutional practices, emphasizing those of EPA to assess environmental agents, and on the types of data generated through regulatory testing strategies, such as those discussed in Chapter 4. The chapter concludes with observations regarding strengths and weaknesses of the current system for generating toxicologic data to assess environmental risks.

---

[1]The term *risk-assessment guideline* is used here instead of *inference guideline*, a term of art defined in NRC (1983) to be the set of principles followed by risk assessors in interpreting and reaching judgments based on scientific data. Risk-assessment guidelines are different from testing guidelines, which provide protocols for specific types of toxicity tests.

NONCANCER RISK-ASSESSMENT GUIDANCE

Neurotoxicity

EPA (1998), the International Programme on Chemical Safety (IPCS) (2001), and the Organisation for Economic Co-operation and Development (OECD) (2004) have provided guidance on the use and interpretation of data generated by neurotoxicity tests. A few of the key principles used to evaluate neurotoxicity data for risk-assessment purposes are highlighted here to show what the regulatory data needs for toxicity testing are and where available data may fall short.

Definition of Neurotoxic Effects

EPA, IPCS, and OECD define *neurotoxicity* as an adverse change in the structure or function of the central or peripheral nervous system after exposure to a chemical, physical, or biologic agent. Adverse effects include alterations from baseline or normal conditions that diminish an organism's ability to survive, reproduce, or adapt to the environment.

As discussed in Chapter 2, neurotoxic effects can be functional or structural. Functional effects are neurochemical, neurophysiological, or behavioral and include adverse changes in somatic, autonomic, sensory, motor, and cognitive function (IPCS 2001). Structural neurotoxic effects are adverse neuroanatomic changes at any level of nervous system organization. Central nervous system (CNS) neurons generally cannot be replaced after damage, so toxic damage to the brain or spinal cord that results in neuronal loss is usually permanent. If axons in peripheral nerves are damaged, they can regenerate and reach their original target site if the neuronal cell bodies are not damaged. Axons in the CNS, most notably in the spinal cord, may also regenerate but are less likely to reach their original targets.

Neurotoxic effects may be either direct or indirect. Direct effects result when agents or their metabolites act directly on sites in the nervous system. Indirect effects result if agents or metabolites produce their effects primarily by interacting with sites outside the nervous system—that is, the effects are secondary to other systemic toxicity. To determine whether neurotoxic effects are direct or indirect, gross toxicity, losses in body weight, and alterations in normal metabolism are evaluated for their possible relationship to the observed effects (IPCS 2001). However, distinguishing between direct and indirect effects may be difficult, particularly when the mechanisms of neurotoxicity are not known. EPA, IPCS, and OECD discuss and provide guidance for study interpretation when neurotoxic effects are found at doses that also cause other systemic toxicity.

EPA Concern Levels

EPA neurotoxicity risk assessment distinguishes among levels of concern on the basis of the magnitude of effect, the duration of exposure, and the reversibility of some neurotoxic effects. In general, there is less concern about effects that are rapidly reversible or transitory—specifically those measured in minutes, hours, or days—and that appear to be associated with the pharmacokinetics of the causative agent and its presence in the body. However, EPA and OECD caution that reversible effects should not be readily dismissed, because reversible changes that occur in the occupational setting or the environment may be of high concern—for example, if a short-acting solvent interferes with operation of heavy equipment in an industrial plant. Also, reversible effects resulting from cell death could require activation of repair capacity that decreases future potential adaptability. That is of special concern for the nervous system because neurons, unlike other cells, do not repair damage to DNA or undergo a continual cycle of programmed cell death and replacement. Clear, demonstrable, irreversible change in either the structure or function of the nervous system causes greater concern. Evidence of progressive, delayed-onset, residual, or latent effects also generates a high level of concern.

Assessment of Neurotoxic End Points

EPA's assessment of commonly measured neurotoxic end points is discussed in this section to illustrate some of the approaches used to evaluate neurotoxicity data. Behavioral end points are measured with a functional observational battery and motor-activity tests. Those tests are designed to detect and measure major overt behavioral, physiologic, and neurologic signs. EPA, IPCS, and OECD guidelines emphasize the importance of evaluating data in terms of patterns of effects rather than as isolated independent end points. There is a potential for false-positive statistical findings because of the large number of end points typically evaluated. Thus, the relevance of statistically significant test results should be evaluated according to the number of signs affected; the pattern of effects with respect to functional domains (such as neuromuscular, sensory, and autonomic); the doses at which effects are observed; the nature, severity, and persistence of the effects; and their incidence compared with that in control animals.

If only a few unrelated measures are affected or the effects are unrelated to dose, the results might not indicate a neurotoxic effect (EPA 1998). If several neurologic signs are affected, but only at the high dose and in conjunction with other overt signs of toxicity, EPA does not consider it to be persuasive evidence of a direct neurotoxic effect. For example, body-weight changes can affect measurements of auditory startle, and temperature can affect conduction velocity. If several related measures in a battery of tests are affected and the effects appear to be dose-dependent, the data are considered to be evidence of a direct neurotoxic effect, especially in the absence of other systemic toxicity. However, the observation of some specific end points, such as body tremors and convulsions, even of short duration and even if they are the only observable changes, may be sufficiently important to raise a high level of concern (OECD 2004).

Tests that measure more complex behaviors—such as tests of schedule-controlled operant behaviour, learning, and memory—often require that the test animals have adequate motivation or intact sensory and motor function. Improved performance of a complex task does not necessarily indicate the absence of neurotoxicity; both increases and decreases in neurobehavioral performance may result from deleterious chemical interactions with neurons (IPCS 2001). Thus, EPA does not consider an improvement to be adverse or beneficial until it is so demonstrated by converging evidence.

Some neurotoxicity-testing protocols suggest using a high dose that produces minimal toxicity because behavioral and functional findings can be difficult to interpret when substantial systemic toxicity is observed. In designing studies that include special behavioral, morphologic, neurochemical, or neurophysiologic measures, OECD recommends that careful consideration be given to doses and study conditions that minimize confounding effects of generalized systemic toxicity.

The development of specific and selective biomarkers of neurotoxicity could theoretically improve assessment of the neurotoxic potential of chemicals. IPCS observed that neurotoxicology lags behind other branches of toxicology in the development of such biomarkers. The lack of progress can be attributed partially to the complexity of the nervous system, the multiplicity of expressions of neurotoxic effects, and the limited understanding of the mechanism of action of many neurotoxic agents (IPCS 2001).

Developmental Neurotoxicity

Although the general principles for evaluating adult neurotoxic end points apply to the developing animal, there are issues of particular importance in the evaluation of animal developmental neurotoxicity studies that affect risk assessment. The development of the mammalian nervous system is a highly complex process with specialized morphologic and biochemical patterns of organogenesis that continue as a carefully timed multistage process guided by chemical messengers (IPCS 2001). A relatively minor disturbance resulting in a perturbation of the developmental interactions between selective cells for a short time may have major effects on the developing CNS. In addition, blood-brain barriers that will eventually protect much of the adult brain, spinal cord, and peripheral nerves are incomplete. As a result, risk assessment of acute and repeated exposures to females of childbearing age (13 years old and older) should include careful consideration of potential exposure of the fetus and its effects on the developing nervous system. Toxicity data can assist in determining whether developmental effects are due primarily to acute or repeated exposures in utero or postnatally.

EPA states that chemical agents that produce developmental neurotoxicity at a dose that is not toxic to the maternal animal are of special concern, whereas EPA generally discounts developmental neurotoxic effects when overt maternal toxicity is moderate or greater. However, EPA cautions that current information is inadequate to assume that developmental effects at doses that cause minimal maternal toxicity result only from maternal toxicity. Another possibility is that both the mother and the developing nervous system are equally sensitive to a given dose. More important, EPA notes that "whether developmental effects are secondary to maternal toxicity or not, the maternal effects may be reversible while the effects on the offspring may be permanent" (EPA 1995).

EPA emphasizes that developmental neurotoxicity should be evaluated in light of other toxicity data, including those on other types of developmental toxicity. Methods of and approaches to toxicity testing that improve the understanding of the mechanisms of neurotoxicity and of the mechanisms by which maternal toxicity and stress can cause structural and functional effects on the developing fetus could lead to improved risk assessment.

Categories and Overall Evaluation of Neurotoxicity Evidence

EPA (1995) guidelines call for summarizing the evidence from the neurotoxicity database into categories of "sufficient evidence," "sufficient human evidence," "sufficient experimental animal evidence/limited human data," and "insufficient evidence." The "sufficient evidence" category includes data that collectively provide enough information to judge whether a human neurotoxic hazard could exist. The "sufficient experimental animal evidence/limited human data" category is used when the evidence is judged to support a conclusion of potential or lack of potential neurotoxic hazard. Strong findings from one guideline study are sufficient to establish potential, whereas findings from more than one study and in multiple species are needed to establish that neurotoxic potential does not exist. EPA, IPCS, and OECD emphasize the importance of evaluating overall patterns of effects relative to other neurotoxicologic measures and systemic toxicity end points to determine level of concern and severity of effect and to address possible confounding by systemic toxicity.

OECD (2004) emphasizes an iterative approach to determining whether experimental neurotoxicity data are sufficient for risk assessment. OECD considers initial neurotoxicology testing to be standard acute and repeated-dose toxicity studies in which functional or histologic information is gathered on all major organ systems, including the nervous system. All available data, including human and animal neurotoxicology data on structurally related chemicals, are assessed. The need for additional studies is based on hazard characterization and exposure assessment to determine whether the available data are sufficient to evaluate risk in light of the intended use, foreseeable misuse, and special considerations of exposed populations, such as sex and age.

Reproductive and Developmental Toxicity

EPA has developed guidelines for developmental-toxicity risk assessment (1991) and reproductive-toxicity risk assessment (1996a). As discussed in Chapter 2, reproductive and developmental toxicity testing

includes a broader category of end points than most kinds of toxicity testing because of the multiple life stages of exposure and the variety of effects that can result. Exposure of sexually mature animals before conception can result in infertility and decreased fertility; exposure during
pregnancy can result in embryonic death, congenital malformations, fetal
growth retardation, and premature or delayed parturition; and exposure
of the neonatal, immature, or adolescent organism may result in death,
growth retardation or stimulation, endocrine abnormalities, immunologic
deficits, neurobehavioral effects, or cancer.

In 1991, EPA published *Guidelines for Developmental Toxicity
Risk Assessment* (1991), which outlines the principles and methods for
evaluating exposure data from animal and human studies to characterize
risks to human development, growth, survival, and function. The EPA
document provides guidance on the relationship between maternal and
developmental toxicity, characterization of the health-related database
for developmental-toxicity risk assessment, use of the benchmark-dose
approach in dose-response assessment, and application of the reference-
dose or reference-concentration approach to developmental-toxicity assessment.

In 1996, EPA published *Guidelines for Reproductive Toxicity Risk Assessment* (1996a). Those guidelines focus on the reproductive-system function as related to sexual behavior, fertility, pregnancy outcomes, lactating
ability, and effects on gametogenesis, gamete maturation and function,
reproductive organs, and components of the endocrine system that directly support reproductive functions. The guidelines concentrate on the
integrity of the male and female reproductive system necessary to ensure
successful procreation. They also emphasize the importance of maintain-
ing the integrity of the reproductive system for overall physical and psy-
chological health.

The guidelines used by EPA, the UN, and OECD are fairly comparable regarding definition and evaluation of end points. However, in the
European Union (EU), the data are used to classify chemicals into three
hazard categories: substances that are known to cause (category 1),
should be regarded as causing (category 2), or cause concern about
(category 3) impairment of fertility or developmental toxicity in humans
(EU 2001). The UN, in its Global Harmonization System of Classification and Labeling of Chemicals (UN 2003), has combined the first two of
the EU categories and formed two main hazard categories: "known or
presumed" (category 1) and "suspect" (category 2) human reproductive
or developmental toxicant (UN 2003). Category 1 is subdivided into
categories of known human reproductive or developmental toxicant

(1A), indicating evidence primarily from human studies, and presumed human reproductive or developmental toxicant (1B), indicating clear evidence primarily from animal experiments. A chemical is in category 1B if animal studies provide clear evidence of toxicity that is not found to be a secondary, nonspecific consequence of other toxic effects. When mechanistic data raise substantial doubt about the relevance of the findings to humans or there is some (but not clear) evidence from animal and human studies, a chemical may be placed in category 2. The EU and UN systems generally require direct evidence from animal or human studies for a chemical to be placed in known, presumed, or suspect categories.

Noncancer Dose-Response Assessment

Risk assessments for end points other than cancer are based on the idea that there is a magnitude of exposure—a threshold—at or below which effects do not occur and above which they do. The rationale and methods for characterizing noncancer dose-response relationships are described in guidance documents by EPA (2004a), IPCS (1999, 2001), the National Research Council (NRC 1977, 2001), and other institutions. According to the EPA (2005a) cancer risk-assessment guidelines, non-cancer dose-response methods apply when carcinogens are judged to have a threshold mode of action.

In practice, it has not been possible to define objective criteria for establishing thresholds (Daston 1993). Thresholds may be biologically plausible for some end points, but the variation in response of sensitive members of the population makes it difficult to determine an absolute threshold for the entire population. Nonlinear, or threshold, mathematical models are generally not used to describe the relationship between risk and dose at low doses quantitatively. Instead, the dose-response relationship is characterized by values, such as a no-observed-adverse-effect level (NOAEL) or a benchmark dose (BMD). NOAELs and BMDs can then serve as the basis for identifying a reference dose (RfD), a reference concentration (RfC), a tolerable intake, or a guidance value—exposure levels at or below which significant adverse effects are not thought to occur (EPA 2004a; IPCS 1994). Occasionally, a nonthreshold dose-response relationship is judged plausible or likely for a noncancer end point, at least at doses to which humans are commonly exposed. When that occurs, different assessment approaches may be applied.

Guidance values, such as RfDs and RfCs, are used by regulatory agencies to establish levels of daily exposure below which no adverse effects would be expected to occur even in potentially sensitive individuals, such as children. Examples include maximum contaminant levels for drinking-water contaminants, ambient-air quality concentrations of air pollutants, acceptable daily intakes of pesticides and other food residues, acute-exposure guideline levels for accidental releases, and other advisories.

The Reference-Dose Method

The toxicity database on a compound typically includes studies that assess a multitude of biologic end points and evaluate its effect on various organ systems. The RfD approach, which is analogous to the approach used by IPCS for developing tolerable intakes, begins by identifying studies of suitable quality and then selecting the most sensitive study from among them. The key study provides dose-response data on the critical effect, which is the most sensitive adverse response that occurs at the lowest dose. The dose-response data are examined to derive or select a "point of departure," which is the starting dose for the calculation of the RfD. The point of departure can be a NOAEL (the highest dose that is not significantly different from the control group) or the lowest-observed-adverse-effect level (LOAEL) (the lowest dose where there is a significant increase compared with the control group). Alternatively, a BMD can be derived. The BMD is the dose that is estimated through model-fitting to produce a specified response rate (for example, 1% or 10%). The BMD is chosen to be in the lower end of the dose-response range for which there are sufficient data. The point of departure is used in conjunction with uncertainty and correction factors (for simplicity, denoted UFs) to derive the RfD (IPCS 1994, 1999; EPA 1991, 1995, 1996a; Dourson 1994).

UFs are used to account for several specific issues, including interspecies extrapolation (UF_A), human intraspecies variability (UF_H), extrapolation between subchronic and chronic exposure durations (UF_S), extrapolation of a LOAEL to a NOAEL if a LOAEL is used (UF_L), and concerns about the quality or breadth of the database (UF_{DB}).

IPCS (1999) assumes factors of 10 for both the interspecies and intraspecies UFs and separates them into pharmacokinetic and pharmacodynamic components (Figure 5-1) when there are sufficient data to derive one of the components (Renwick 2000; Renwick and Lazarus 1998). For example, the pharmacokinetics of interspecies differences may be sufficiently well understood to build and assign parameters to a physiologically based pharmacokinetic model. Model predictions of interspe-

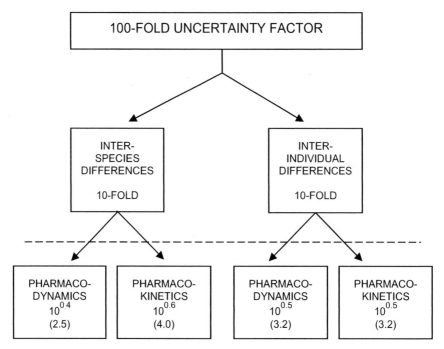

FIGURE 5-1 IPCS subdivision of interspecies and intraspecies uncertainty factors into pharmacokinetic and pharmacodynamic components. Source: IPCS 1999. Reprinted by permission of the International Programme of Chemical Safety, copyright 1999.

cies differences may replace the pharmacokinetic component of the interspecies factor. The remaining uncertainty in interspecies pharmacodynamic differences would be addressed by an interspecies pharmacodynamic factor adopted by IPCS to be 2.5. EPA (2002a) has recognized the IPCS approach and now uses chemical-specific adjustment factors instead of default values when it finds the available data sufficient to derive them.

Individual UFs—usually with values of 10, 3, or 1—are multiplied together to produce an aggregate or composite UF. When the composite UF is large, EPA may judge the data to be insufficient for establishing an RfD and refrains from setting one. The agency has not used a composite UF exceeding 3,000 in over a decade, and an EPA technical panel has recommended that it become policy not to do so (EPA 2002a). EPA (2002a, 2002b) has reviewed the extent to which the composite UF results in adequate protection.

When animal data are used to derive the RfD, the BMD (or the NOAEL or LOAEL) is first adjusted to represent a daily continuous exposure of a human population. For example, the adjustment factor (ADJ) might be 5/7 to adjust a laboratory exposure of 5 days/week to an exposure of 7 days/week. The resulting value is divided by the composite uncertainty, as illustrated in Equation 1.

$$RfD = [(BMD)(ADJ)]/ UF_{composite}] \qquad (1)$$

The Reference-Concentration Method

Guidelines for determining inhalation RfCs have been refined to base the derivation on dose delivered to epithelial regions of the respiratory tract (EPA 1994). In a procedure similar to that for RfD derivation, the RfC is calculated from a benchmark concentration (BMC) or a no-observed-adverse-effect concentration. It is first adjusted to represent a daily continuous exposure and then corrected with dosimetric adjustment factors (DAFs) to provide a human equivalent benchmark concentration ($BMC_{(HEC)}$). The DAFs are based on general knowledge of particle and regional gas deposition characteristics in the specific regions of the respiratory tract in the test animals and humans. The $BMC_{(HEC)}$ is estimated from the following relationship:

$$BMC_{(HEC)} = (BMC_{(Adjusted)})(DAF) \qquad (2)$$

The DAF correction is to estimate an equivalent deposition in the human respiratory tract at the animal $BMD_{(Adjusted)}$. After the "above-the-line" corrections in Equation 2, UFs are applied. Because the $BMC_{(HEC)}$ includes an adjustment for delivered dose between species, EPA applies a UF_A of 3.15 rather than 10, where 3.15 is about the square root of 10. The UF_A is reduced because the correction with the DAF is intended to account for species pharmacokinetic differences but does not account for pharmacodynamic uncertainties in interspecies extrapolation.

Uncertainty Factor for Susceptibility of the Fetus and the Young— Application of the Food Quality Protection Act

The approach of dividing the dose at which responses are observed in animals by some factor or group of factors began in the 1950s when FDA used a factor of 100 to determine allowable human daily intakes of food additives and other compounds on the basis of animal studies

(Lehman and Fitzhugh 1954). In the first application, the adjustments were considered to be safety factors. The major grouping of current UFs discussed above was developed in the 1980s as EPA began a process of standardizing the risk-assessment approach for end points other than cancer (Dourson and Stara 1983; Dourson 1994). In 1993, NRC addressed the question of whether the regulatory approaches for controlling pesticide residues in foods adequately protected infants and children (NRC 1993). After reflecting on the testing and standard-setting system, NRC recommended that a UF of up to 10 be applied when there is evidence of fetal and postnatal developmental toxicity and when data from toxicity testing relative to children are incomplete. Its report stated that "the committee wishes to emphasize that this is not a new, additional uncertainty factor but, rather, an extended application of an uncertainty factor now routinely used by the agencies for a narrower purpose [to address fetal developmental toxicity]. In the absence of data to the contrary, there should be a presumption of greater toxicity to infants and children." The FQPA, signed into law in August 1996, directed EPA to use an additional 10-fold margin of safety in setting pesticide tolerances for infants and children to take into account the potential for prenatal and postnatal toxicity and the completeness of the toxicology and exposure databases. The law provides for departure from the 10-fold margin when reliable evidence shows that a different margin is protective of infants and children [Section 408(b)(2)(C)].

In implementing the FQPA, the EPA Office of Pesticide Programs considers the FQPA factor as part of the tolerance evaluation. The toxicology and exposure data are evaluated, and then it is determined whether there are important residual uncertainties regarding potential risks to the young. EPA uses a weight-of-evidence approach to judge the degree of concern for potential prenatal and postnatal toxicity to determine whether the 10-fold margin, referred to by EPA as the 10X factor, should be used or a different value should be assigned. The approach considers several factors, including available human and animal data on prenatal and postnatal toxicity, the nature of the dose-response relationship, and information on the human relevance of data from animal experiments, such as pharmacokinetics, mechanism of action, and similarity of the biologic response in different species. Table 5-1 illustrates how EPA may weigh those factors in evaluating the necessary FQPA margin.

In practice, EPA's Office of Pesticide Programs usually has found that application of the traditional factors provides adequate protection.

TABLE 5-1 EPA's Weight-of-Evidence Approach for Evaluating Degree of Concern for Prenatal and Postnatal Toxicity on the Basis of Human and Animal Data

Factor	Degree of Concern	
	Increasing Weight—Higher Degree of Concern	Decreasing Weight—Lower Degree of Concern
Pre- and postnatal toxicity	Effects found in humans related to exposure Same types of effects seen in more then one species Effects of a different type with greater potential consequences in young than in adults Persistence or relatively longer recovery of effects in young than in adults	No adverse human or animal effects associated with exposure Similar response in young with relatively shorter recovery than in adults
Dose-response relationship	Effects observed at a lower dose in young than in adults NOAEL not identified Poor data on dose-response relationship	Effects at higher dose in young than in adults or only at high dose in presence of severe generalized toxicity Good data on dose-response relationship that allows confident identification of NOAEL or BMD
Pharmacokinetics	Metabolic profile indicates higher internal dose of active moiety in young than in adults or in humans than in animals	Metabolic profile indicates lower internal dose of active moiety in young than in adults or in humans than in animals
Mode of action	Mode of action supports relevance to humans and concern for animal findings Mode of action may lead to several adverse consequences in offspring	Evidence indicates that mode of action is species-specific and thus not relevant to humans Evidence indicates that humans are less sensitive than animal model

Source: Adapted from EPA 2002a.

Pesticide Fact Sheets summarizing many pesticide evaluations conducted since the passage of the FQPA in 1996 are posted on EPA's web site. Of the 59 chemicals posted, EPA found it unnecessary to apply an FQPA factor—that is, it uses a factor of 1—for all but 11 chemicals. For five pesticides, the agency applied the full FQPA factor of 10 for at least one exposure group and exposure circumstance, such as acute dietary exposure of women of childbearing age. For six pesticides, EPA applied an FQPA factor of 3. In the five cases where the factor of 10 was applied, severe developmental toxicity end points, such as multiple malformations and fetal death, were observed in laboratory animals. In two cases in which an FQPA factor of 1 was applied, a database UF of 10 was used. Both cases were driven by findings from existing studies suggesting effects at lower doses and the need for specific studies to resolve the uncertainty (for example, a developmental-immunotoxicology test).

EPA (2002b) notes that there should be consistency among agency programs, including the pesticide program, in deriving RfDs and RfCs for the same chemical. EPA (2002b) found that broad use of the database UF under the FQPA is characteristic of other agency practices and provided the following guidance on its use in risk assessments:

> The [database] "completeness" inquiry should be a broad one that takes into account all data deficiencies. In other words, the risk assessor should consider the need for traditional uncertainty factors not only when there are inadequacies or gaps in currently required studies on pesticides, but also when other important data needed to evaluate potential risks to children are missing or are inadequate.

EPA (2002b) also noted that all agency programs have traditionally considered a group of five studies to be the minimum for deriving a "high-confidence" chronic RfD—two chronic oral studies in different species, two prenatal-development studies in different species, and a multigeneration reproductive-toxicity study in rats. EPA concluded that

> the absence of any of these studies suggests that the existing data are not sufficient to address and relieve uncertainties regarding the hazards of the chemical and would typically give rise to the need for a database uncertainty factor to protect the safety of infants and children.

In addition to considering any data gaps involving these five studies, the risk assessor should as is now standard Agency practice evaluate other data gaps, particularly those that pertain to evaluating risk to children and other sensitive subpopulations.

However, EPA (2002b) continued that when determining the need for a database UF, the risk assessor should evaluate how likely the missing or inadequate study will substantially change the outcome of the overall risk assessment. EPA (2002a) found that when the traditional UFs are appropriately applied, they are usually adequate and that an additional factor, such as an FQPA factor, was not needed to address concerns regarding prenatal and postnatal toxicity.

EPA (2002b) is considering new studies and modification of existing guideline studies to provide a more comprehensive coverage of life stages, a more systematic evaluation of pharmacokinetics, and a more focused evaluation of structural and functional toxicity in the young. Such studies might include pharmacokinetics in fetuses or young animals, direct dosing of offspring before weaning, enhanced developmental-neurotoxicity studies, developmental-immunotoxicity studies, and enhanced evaluations related to endocrine disruption.

CANCER RISK-ASSESSMENT GUIDANCE

Principles for assessing cancer hazards and risks have been in use at least since the early 1970s (OTA 1987). In the early 1980s, NRC recommended periodic updating of risk-assessment guidelines to keep pace with scientific advances and to clarify science policy positions (NRC 1983). Several agencies and organizations have developed and implemented guidelines or principles for cancer risk assessment. EPA (1986, 1996b, 1999, 2005a) has developed, revised, and conducted peer reviews of its cancer risk-assessment guidelines over the years as the scientific basis of evaluation has evolved. The current EPA (2005a) carcinogen guidelines emphasize cancer hazard identification and dose-response assessment and provide limited guidance for carcinogen exposure assessment and risk characterization. EPA (2005b) has also published supplemental guidelines to describe possible approaches for assessing risks resulting from early-in-life exposures to carcinogens.

Cancer Hazard Identification

Carcinogen hazard-identification guidelines provide approaches for evaluating data to determine a chemical's cancer-causing potential. The National Toxicology Program (NTP 2005) has general guidance to identify a chemical "known to cause cancer" or "reasonably anticipated to be a human carcinogen." The International Agency for Research on Cancer (IARC 1999) has developed more detailed guidance to categorize a chemical as a known, probable, or possible human carcinogen or as a chemical for which inadequate evidence is available or for which evidence suggests lack of carcinogenicity. Similar classifications have been adopted by EPA (2005a) and the Institute of Medicine (IOM 2004).

Classification systems typically involve two steps. First, the different types of evidence on a chemical are evaluated for sufficiency for establishing causal relationships between cancer and exposure to the chemical. Second, there is an overall evaluation of the entire body of evidence.

There are three types of evidence. There is human evidence from cancer epidemiology studies; bias, confounding, and chance are critically evaluated to determine the extent to which they might explain observed relationships. There is evidence from animal cancer bioassays; finding the evidence sufficient to establish causality requires multiple studies showing increases in cancers or tumors that can progress to cancer. And there are other relevant data, such as "data on preneoplastic lesions, tumour pathology, genetic and related effects, structure-activity relationships, metabolism and pharmacokinetics, physicochemical parameters and analogous biological agents" (IARC 1999). Data are "considered to be especially relevant if they show that the agent in question has caused changes in exposed humans that are on the causal pathway to carcinogenesis" (IARC 1999).

Table 5-2 shows how the overall evidence from human, animal, and other relevant studies may be used by various agencies to reach conclusions about potential carcinogenicity. Positive categorizations of carcinogenicity—such as "known," "sufficient," "likely," and "suggestive"—require, at a minimum, direct observations of cancer in humans or laboratory animals. Rarely, structural analogy to an established carcinogen may be used. IOM has the most stringent criteria and requires epidemiologic evidence for drawing any positive conclusions about potential carcinogenicity; animal evidence and other test information are used only to confirm cancer causation once epidemiologic associations have

TABLE 5-2 Level of Evidence in Carcinogen Classification Schemes or Narrative Descriptions

Type of Evidence	IARC	EPA	NTP	IOM
Sufficient human, sufficient animal	Carcinogenic to humans	Carcinogenic to humans	Known to be human carcinogen	Sufficient evidence of causal relationship
Sufficient human	Carcinogenic to humans	Carcinogenic to humans	Known to be human carcinogen	Sufficient evidence of association
Limited human, sufficient animal, strong evidence in exposed humans that agent acts through relevant mechanism of carcinogenicity	Carcinogenic to humans	Carcinogenic to humans	Known to be human carcinogen	Limited/suggestive evidence of association
Limited human, sufficient animal	Probably carcinogenic to humans	Likely to be carcinogenic to humans	Reasonably expected to be human carcinogen	Limited/suggestive evidence of association
Inadequate human, sufficient animal, strong evidence that carcinogenesis is mediated by a mechanism that also operates in humans	Probably carcinogenic to humans	Likely to be carcinogenic to humans	Reasonably expected to be human carcinogen	Inadequate/insufficient evidence to determine whether association exists
Inadequate human, limited animal, strong supporting evidence from other relevant data	Possibly carcinogenic to humans	Likely to be carcinogenic to humans	Reasonably expected to be human carcinogen	Inadequate/insufficient evidence to determine whether association exists

(Continued)

TABLE 5-2 *Continued*

Type of Evidence	IARC	EPA	NTP	IOM
Inadequate human, sufficient animal	Possibly carcinogenic to humans	Likely to be carcinogenic to humans	Reasonably expected to be human carcinogen	Inadequate/insufficient evidence to determine whether association exists
Limited human, limited or inadequate animal	Possibly carcinogenic to humans	Suggestive evidence of carcinogenic potential	Reasonably expected to be human carcinogen	Limited/suggestive evidence of association
Inadequate human, limited animal	Not classifiable as to carcinogenicity in humans	Suggestive evidence of carcinogenicity potential	(No statement)	Inadequate/insufficient evidence to determine whether association exists
Inadequate human, inadequate animal, convincing relevant information that the agent acts through mechanisms indicating it would likely cause cancer in humans	Not classifiable as to carcinogenicity to humans	Inadequate evidence to assess carcinogenic potential	Reasonably expected to be human carcinogen	Inadequate/insufficient evidence to determine whether association exists
Sufficient animal, strong evidence that mechanism of carcinogenicity in animals does not operate in humans	Not classifiable as to carcinogenicity in humans	Inadequate evidence to assess carcinogenic potential	(No statement)	Inadequate/insufficient evidence to determine whether association exists

Evidence suggesting lack of carcinogenicity in humans and experimental animals or evidence suggesting lack of carcinogenicity in experimental animals consistently and strongly supported by broad range of other relevant data	Probably not carcinogenic in humans	Not likely to be carcinogenic in humans	(No statement)	Inadequate/ insufficient evidence to determine whether association exists

Sources: IARC 1999; IOM 2004; EPA 2005a; NTP 2005.

been demonstrated. IARC and EPA use direct evidence of cancer from animal bioassays to determine whether a chemical may or is likely to cause cancer. IARC requires more evidence for a conclusion of probable carcinogenicity than EPA needs for a conclusion of likely carcinogenicity. Multiple studies are usually required to establish a positive categorization. IARC (2005b), however, is now considering modifying its rules of evidence so that "possible carcinogenicity can be assessed solely on the basis of strong evidence from mechanistic and other relevant data." Current NTP guidance indicates that an agent can be classified as reasonably expected to be a human carcinogen when there is "convincing relevant information that the agent acts through mechanisms indicating it would likely cause cancer in humans" (NTP 2005). In practice, however, this criterion has not been used in the absence of direct evidence of carcinogenicity of a chemical or a closely related structural analogue.

The hazard-identification guidelines of IARC and EPA discuss in detail how some design features of the bioassay may influence inferences. EPA's guidelines (2005a) note that study findings can be compromised by dose selection. High doses that cause excessive toxicity complicate the interpretation of tumor observations. Doses that are set too low render a study insensitive. Too few doses or too few animals at each dose limit the dose-response characterization. Guidelines also discuss study quality, reporting, and interpretation with regard to statistical and biologic significance, use of historical and concurrent control animals, and tumor type and progression.

In the IARC, EPA, and NTP guidance, indirect evidence is used to affect the categorization that would otherwise be based on the direct evidence alone. Indirect evidence from genotoxicity assays, comparative human and animal metabolism and receptor profiles, and structure-activity, biomarker, and other studies may increase the confidence that cancer findings in animals are relevant or irrelevant to humans. Such data have been used to classify some chemicals (for example, dioxin and ethylene oxide) as carcinogenic to humans in the absence of definitive epidemiologic data. A potential for carcinogenicity should not be based on indirect evidence in the absence of direct findings of cancer in animal or human studies.

Cancer Dose-Response Assessment

In the EPA (2005a) guidelines, mode-of-action data guide the dose-response assessment. A two-step process is used: the first step involves the determination of mode of action for each tumor finding, and that dictates the approach for the dose-response analysis, which may be a linear analysis that presumes a linear dose-response relationship, a nonlinear analysis that reflects the assumption of a threshold, or both. When there is strong evidence of genotoxicity from multiple test systems, a linear relationship is assumed. A nonlinear mode of action is determined from data usually at the cellular, tissue, organ, and organism level that together indicate that exposures at some dose would be without cancer effect. The guidelines provide general criteria for the evaluation of data in assessing the mode of action.

In a linear analysis, mathematical models are fitted to dose-response data to estimate a benchmark dose. The BMD is a dose that causes a specified fraction of subjects to develop tumors. The BMD is then used to infer lower risk-specific doses. In a threshold, nonlinear analysis, the standard approach used for setting an RfD for noncancer end points, as discussed above, is used.

The EPA (2005a) guidelines provide for the application of a biologically based model for agents on which quantitative data relate specific key events in the cancer process to neoplasia. A large amount of data is required to support biologically based modeling. Standards and guidance for data generation are not available, nor is specific guidance available for the use of such data in dose-response evaluation.

The EPA guidelines note the importance of considering potentially susceptible populations, such as children and other "subpopulations of individuals who are particularly vulnerable to the effects of an agent because of pharmacokinetic or metabolic differences (genetically or environmentally determined)" (Bois et al. 1995). In practice, few assessments quantitatively characterize human variability in cancer risk. There is a large degree of heterogeneity among humans compared with the relative homogeneity of bioassay animals used as the basis for many cancer dose-response characterizations.

There is no systematic approach for identifying sensitive populations in conducting cancer risk assessment. Human and animal studies indicate that the young can be (but are not always) more sensitive than adults. EPA has developed guidance for assessing the contribution of early life exposures to lifetime cancer risk (EPA 2005b).

The EPA guidelines discuss in many places the issue of cross-species site concordance—that is, whether the specific tumors observed in an animal bioassay should also be assumed to occur in humans. EPA (2005a) states that site concordance of tumor effects between animals and humans should be considered in each case. Thus far, there is evidence that growth control mechanisms at the level of the cell are homologous among mammals, but there is no evidence that these mechanisms are site concordant. Moreover, agents observed to produce tumors in both humans and animals have produced tumors either at the same (e.g., vinyl chloride) or different sites (e.g., benzene) (NRC, 1994). Hence, site concordance is not assumed between animals and humans. On the other hand, certain processes with consequences for particular tissue sites (e.g., disruption of thyroid function) may lead to an anticipation of site concordance.

Although that is EPA's stated position, in many dose-response analyses that use pharmacokinetic information, site concordance has been assumed. There is no clear guidance on when it is appropriate to assume it.

COMMITTEE OBSERVATIONS CONCERNING TOXICITY DATA AVAILABLE FOR RISK ASSESSMENT

Guidelines for assessing hazards and dose-response relationships from toxicologic and epidemiologic data have coevolved with scientific developments and laboratory capabilities. In some respects, the human and animal data being generated as described in Chapters 2 and 3 mesh well with the evidence requirements. In other respects, there is a disconnect between the data needed for evaluating risk and the data generated in the laboratory or field. The following discussion presents the committee observations on data availability and needs for risk assessment.

Coverage of End Points

For widely used drugs, food additives, and pesticides, there is a reasonably good basis for risk-based decision-making. That is not the case for industrial chemicals, partly because there is no mandatory testing of industrial chemicals; and the rules of evidence applied in assessing the hazards and risks they pose do not always relate well to the test data being generated. For example, although adequate cancer bioassay and epidemiologic data are not available on many chemicals, short-term test data and structural alerts are. Indeed, in some testing strategies, carcinogen bioassays are not performed or infrequently performed, and genotoxicity data are generated to provide presumptive evidence of carcinogenicity. Under EPA and IARC carcinogen guidelines, direct evidence of cancer in animals or humans is required if a chemical is to be identified as having carcinogenic potential. In practice, when such data are not available, the chemical is classified as having, for example, "inadequate information to assess carcinogenic potential"; cancer risk is not estimated; and the chemical is generally treated as posing zero cancer risk. A system for using indirect evidence from emerging test strategies and genotoxicity batteries could be developed to guide the assessment of chemicals that lack adequate cancer bioassay or epidemiologic data. Similarly, systems and guidance could be created for identifying a potential for neurotoxicity, developmental toxicity, and other toxicities on the basis of short-term tests and high-throughput approaches using end points that are more specific to the relevant toxicologic processes that are conserved across species.

Coverage of Life Stages

Most toxicologic tests do not provide sufficient information to assess health risks associated with exposures at different life stages, including pregnancy, infancy, childhood, and old age. To characterize health effects of exposures at different life stages fully, more elaborate study designs would be needed. The extent to which existing risk-assessment procedures for establishing guidance levels may adequately address health protection in relation to exposures at different life stages is an issue of current scientific inquiry and discussion.

Development of Epidemiologic Evidence

High-quality human evidence is given the most weight in hazard identification. Existing regulatory programs and data-generation requirements do not encourage the development of epidemiologic data, but they could. Followup studies long enough to identify carcinogenic hazards are not now required after the introduction of pharmaceuticals, pesticides, or other chemicals.

Use of Standardized Toxicity Tests

Standardization of animal toxicity tests (for example, standardized as to species and strain) provides stability and predictability in testing and regulatory processes but appears in some cases to work against the development of findings of greatest relevance to humans. The finding of lack of relevance of results does not usually prompt explorations with alternative animal models that may be more relevant. When specific animal findings are not relevant to humans, the lack of additional testing in a more appropriate species biases the process toward the creation of false negatives. However, adherence to the standard study-design features and the current testing guidelines generally does produce data that are valuable for hazard identification. The high doses used in the toxicity tests limit their applicability for characterizing risks at the low doses typical of environmental exposures, but accurate characterizations at the low doses would require impracticably large numbers of animals.

The issue of indirect, systemic toxicity resulting from high-dose testing and the challenges it poses for interpreting findings of cancer, reproductive toxicity, developmental toxicity, and neurotoxicity is discussed in risk-assessment guidelines for these end points. For example, neurotoxicity testing currently relies on apical tests that have a strong emphasis on behavioral end points that can be confounded by other systemic toxicity, such as can be seen at or above the maximum tolerated dose or indicated by moderate maternal toxicity. The development of specific and selective biomarkers of toxicity could theoretically improve assessment of such effects when they occur in the presence of systemic toxicity.

Default Dose-Response Assessments

Regardless of the specific noncancer end point, the dose-response relationship is typically characterized with the use of reference doses and reference concentrations. The starting point for analysis is the selection of the point of departure—a NOAEL, LOAEL, or BMD. Use of toxicity data for establishing a point of departure is often not taken into account in designing experiments. Attention to the number of animals, magnitude of dose, number of dose groups, and dose spacing can lead to NOAELs that fall closer to the actual no-effect levels or more reliable estimates of the BMD. For noncancer end points, a reference level is derived from a

BMD, LOAEL, or NOAEL for the most sensitive adverse end point and specified uncertainty and adjustment factors; and, unless scientific information indicates that it is inappropriate, the same process is applied for carcinogens that are judged to act through a nonmutagenic mechanism. The uncertainty and adjustment factors reflect a variety of considerations, including human heterogeneity, interspecies differences, and completeness of the available data. Some time has passed since the currently used values were adopted. Comprehensive testing and data-collection strategies are needed to re-evaluate those factors or alternatives to their use, such as probabilistic risk assessment for noncancer end points; a general framework for evaluating factors used for developmental-toxicity assessment has been described (NRC 2000).

For mutagenic carcinogens or carcinogens of unknown mechanism, a linear no-threshold dose-response model is used by EPA for low-dose extrapolation. Risk-assessment guidelines for assessing the dose-response relationship of mutagenic carcinogens from animal data assume that each individual faces the same risk of cancer at a given dose. Noncancer guidance applies a generic default factor to adjust for variability. Testing strategies do not reflect a systematic approach for developing data to assess the variability of human responses to chemicals quantitatively. Such data would aid in understanding whether the current default procedures for estimating cancer risk are conservative overall or may in some cases understate the risk for some segments of the population.

Data and Framework for Nondefault Assessment

For most environmental agents of concern, the initial default assessment of risk involves extrapolation of findings from studies in very small homogeneous animal populations that are exposed for a portion of their lifespan at doses typically considerably higher than environmental levels to large heterogeneous human populations. The extrapolations have the potential to overestimate or underestimate risk; when the difference between expected human exposure and effect level is relatively small or the costs of indicated exposure reductions are high, additional study may be undertaken either by regulatory authorities or by affected industries.

With the exception of standard genotoxicity testing, the generation of data for mode-of-action evaluations, pharmacokinetic modeling, and other nondefault approaches is typically ad hoc. The data may be supplied by interested parties or otherwise available in the literature but are not required by the agencies. Although the guidelines may provide a loose framework for nondefault approaches, they provide little guidance on data-generation issues, such as hypothesis-testing of modes of action and plausible competing hypotheses. Guidance is also limited or non-existent for developing other information useful for nondefault analyses, including data for models for assessing human variability, age dependence, site concordance, and high- to low-dose and cross-species differences in pharmacokinetics.

Optimizing further testing to improve the initial characterization of a particular chemical or class of chemicals can be highly context-dependent. Nonetheless, a general framework and further guidance on developing a testing strategy to improve specific risk assessments would be useful. In the process for setting national ambient air quality standards (NAAQS) for criteria air pollutants, data generation, assessment, and standard-setting itself are components of an iterative and cyclic process. As EPA is adopting the new standard, the stage is set for further study and generation of information to improve assessment in the next round of standard-setting. Formal reviews, such as NRC (2004), can be part of the process that leads to coherent and effective testing strategies.

REFERENCES

Bois, F.Y., G. Krowech, and L. Zeise. 1995. Modeling human interindividual variability in metabolism and risk: The example of 4-aminobiphenyl. Risk Anal. 15(2):205-213.

Daston, G.P. 1993. Do thresholds exist for developmental toxicants? A review of the theoretical and experimental evidence. Pp. 169-197 in Issues and Reviews in Teratology, Vol. 6, H. Kalter, ed. New York: Plenum Press.

Dourson, M.L. 1994. Methods for establishing oral reference doses (RfDs). Pp. 51-61 in Risk Assessment of Essential Elements, W. Mertz, C.O. Abernathy, and S.S. Olin, eds. Washington, DC: ILSI Press.

Dourson, M.L., and J.F. Stara. 1983. Regulatory history and experimental support of uncertainty (safety) factors. Regul. Toxicol. Pharmacol. 3(3):224-238.

EPA (U.S. Environmental Protection Agency). 1986. Guidelines for Carcinogen Risk Assessment. EPA/630/R-00/004. Risk Assessment Forum, U.S. Environmental Protection Agency, Washington, DC [online]. Available: http://www.epa.gov/ncea/raf/car2sab/guidelines_1986.pdf   [accessed March 30, 2005].

EPA (U.S. Environmental Protection Agency). 1991. Guidelines for Developmental Toxicity Risk Assessment. EPA/600/FR-91/001. Risk Assessment Forum, U.S. Environmental Protection Agency, Washington, DC [online]. Available: http://www.epa.gov/ncea/raf/pdfs/devtox.pdf [accessed March 30, 2005].

EPA (U.S. Environmental Protection Agency). 1994. Methods for Derivation of Inhalation Reference Concentrations and Application of Inhalation Dosimetry. EPA/600/8-90/066F. Environmental Criteria and Assessment Office, Office of Health and Environmental Assessment, Office of Research and Development, U.S. Environmental Protection Agency, Research Triangle Park, NC. October 1994.

EPA (U.S. Environmental Protection Agency). 1995. Guidelines for Neurotoxicity Risk Assessment. EPA/630/R-95/001F. Risk Assessment Forum, U.S. Environmental Protection Agency, Washington, DC [online]. Available: www.epa.gov/ncea/raf/pdfs/neurotox.pdf [accessed March 30, 2005].

EPA (U.S. Environmental Protection Agency). 1996a. Guidelines for Reproductive Toxicity Risk Assessment. EPA/630/R-96/009. Risk Assessment Forum, U.S. Environmental Protection Agency, Washington, DC [online]. Available: http://www.epa.gov/ncea/raf/pdfs/repro51.pdf [accessed March 30, 2005].

EPA (United States Environmental Protection Agency). 1996b. Proposed Guidelines for Carcinogen Risk Assessment. EPA/600/P-92/003C. Office of Research and Development, U.S. Environmental Protection Agency, Washington, DC [online]. Available: http://cfpub.epa.gov/ncea/raf/cra_prop.cfm [accessed April 4, 2005].

EPA (U.S. Environmental Protection Agency). 1998. Health Effects Test Guidelines OPPTS 870.6300. Developmental Neurotoxicity Study. EPA 712-C-98-239. Office of Prevention, Pesticides, and Toxic Substances, U.S. Environmental Protection Agency, Washington, DC [online]. Available: http://www.epa.gov/opptsfrs/OPPTS_Harmonized/870_Health_Effects_Test_Guidelines/Series/870-6300.pdf [accessed March 15, 2005].

EPA (U.S. Environmental Protection Agency). 1999. Guidelines for Carcinogen Risk Assessment. Review Draft NCEA-F-0644. Risk Assessment Forum, U.S. Environmental Protection Agency, Washington, DC [online]. Available: http://www.epa.gov/ttn/atw/toxsource/carcinogens.html [accessed March 30, 2005].

EPA (U.S. Environmental Protection Agency). 2002a. A Review of the Reference Dose and Reference Concentration Processes. Final Report. EPA/630/P-02/002F. Risk Assessment Forum, U.S. Environmental Protection Agency, Washington, DC [online]. Available: http://www.epa.gov/iris/RFD_FINAL%5B1%5D.pdf [accessed March 11, 2005].

EPA (U.S. Environmental Protection Agency). 2002b. Determination of the Appropriate FQPA Safety Factor(s) in Tolerance Assessment. Office of Pesticide Programs, U.S. Environmental Protection Agency, Washington,

DC. February 28, 2002 [online]. Available: http://www.epa.gov/oppfead1/ trac/science/determ.pdf [accessed March 30, 2005].

EPA (U.S. Environmental Protection Agency). 2004a. Risk Assessment for Non-cancer Effects. Air Toxics Website. U.S. Environmental Protection Agency [online]. Available: http://www.epa.gov/ttn/atw/toxsource/ noncarcinogens.html [accessed April 1, 2005].

EPA (U.S. Environmental Protection Agency). 2004b. An Examination of EPA Risk Assessment Principles and Practice, Staff Paper Prepared for the U.S. Environmental Protection Agency by Members of the Risk Assessment Task Force. EPA/100/B-04/001. Office of the Science Advisor, U.S. Environmental Protection Agency, Washington, DC [online]. Available: http://www.epa.gov/OSA/ratf-final.pdf [accessed March 22, 2005].

EPA (U.S. Environmental Protection Agency). 2005a. Guidelines for Carcinogen Risk Assessment. EPA/630/P-03/001B. Risk Assessment Forum, U.S. Environmental Protection Agency, Washington, DC. March 2005 [online]. Available: http://www.epa.gov/iris/cancer032505.pdf [accessed Nov. 10, 2005].

EPA (U.S. Environmental Protection Agency). 2005b. Supplemental Guidance for Assessing Susceptibility for Early-Life Exposure to Carcinogens. EPA/630/R-03/003F. Risk Assessment Forum, U.S. Environmental Protection Agency, Washington, DC. March 2005 [online]. Available: http:// www.epa.gov/iris/children032505.pdf [accessed Nov. 10, 2005].

EU (European Union). 2001. European Commission Directives: 67/548/EEC, updated 2001/59/EC. Official Journal of European Communities 225:1-333 [online]. Available: http://europa.eu.int/eur-lex/pri/en/oj/dat/2001/ l_225/l_22520010821en00010333.pdf [accessed Nov. 10, 2005].

IARC (International Agency for Research on Cancer). 1999. Preamble to the IARC Monographs. 12. Evaluation. IARC Monographs Programme on the Evaluation of Carcinogenic Risk to Humans [online]. Available: http://monographs.iarc.fr/monoeval/eval.html [accessed April 4, 2005].

IARC (International Agency for Research on Cancer). 2005a. IARC Monographs Programme on the Evaluation of Carcinogenic Risk to Humans [online]. Available: http://monographs.iarc.fr/ [accessed March 30, 2005].

IARC (International Agency for Research on Cancer). 2005b. DRAFT Preamble to the IARC Monographs. IARC Monographs on the Evaluation of Carcinogenic Risks to Humans. International Agency for Research on Cancer, World Health Organization, Lyon, France [online]. Available: http://monographs.iarc.fr/past&future/PreambleDraft2005.pdf. [accessed Oct. 28, 2005].

IOM (Institute of Medicine). 2004. Gulf War and Health: Updated Literature Review of Sarin. Washington, DC: The National Academies Press.

IPCS (International Programme on Chemical Safety). 1994. Assessing Human Health Risks of Chemicals: Derivation of Guidance Values for Health-Based Exposure Limits. Environmental Health Criteria 170. Geneva:

World Health Organization [online]. Available: http://www.inchem.org/documents/ehc/ehc/ehc170.htm [accessed March 31, 2005].

IPCS (International Programme on Chemical Safety). 1999. Principles for the Assessment of Risk to Human Health from Exposure to Chemicals. Environmental Health Criteria 210. Geneva: World Health Organization [online]. Available: http://www.inchem.org/documents/ehc/ehc/ehc210.htm [accessed March 30, 2005].

IPCS (International Programme on Chemical Safety). 2001. Guidance Document for the Use of Data in Development of Chemical-Specific Adjustment Factors (CSAFs) for Interspecies Differences and Human Variability in Dose/Concentration–Response Assessment. WHO/PCS/01.4. Geneva: World Health Organization. July 2001 [online]. Available: http://www.who.int/ipcs/publications/methods/harmonization/en/csafs_guidance_doc.pdf [accessed March 31, 2005].

Lehman, A.J., and O.G. Fitzhugh. 1954. 100-fold margin of safety. Assoc. Food Drug Off. U.S.Q. Bull. 18(1):33-35.

Lowrence, W. 1976. Of Acceptable Risk: Science and the Determination of Safety. Los Altos, CA: W. Kaufman.

NRC (National Research Council). 1977. Drinking Water and Health. Washington, DC: National Academy of Science.

NRC (National Research Council). 1983. Risk Assessment in the Federal Government: Managing the Process. Washington, DC: National Academy Press.

NRC (National Research Council). 1993. Pesticides in the Diets of Infants and Children. Washington, DC: National Academy Press.

NRC (National Research Council). 1994. Science and Judgment in Risk Assessment Washington, DC: National Academy Press.

NRC (National Research Council). 2000. Acute Exposure Guideline Levels for Selected Airborne Chemicals, Vol. 1. Washington, DC: National Academy Press.

NRC (National Research Council). 2001. Standing Operating Procedures for Developing Acute Exposure Guideline Levels for Hazardous Chemicals. Washington, DC: National Academy Press.

NRC (National Research Council). 2004. Research Priorities for Airborne Particulate Matter. IV. Continuing Research Progress. Washington, DC: The National Academies Press.

NTP (National Toxicology Program). 2005. Report on Carcinogens, 11th Ed. National Toxicology Program, Research Triangle Park, NC [online]. Available: http://ntp.niehs.nih.gov/ntpweb/index.cfm?objectid=035E5806-F735-FE81-FF769DFE5509AF0A [accessed April 4, 2005].

OECD (Organisation for Economic Cooperation and Development). 2004. Guideline Document for Neurotoxicity Testing. ENV/JM/MONO(2004)25. OECD Series on Testing and Assessment No. 20. Organisation for Economic Cooperation and Development, Paris [online]. Available:

http://appli1.oecd.org/olis/2004doc.nsf/linkto/env-jm-mono(2004)25 [accessed April 1, 2005].

OTA (Office of Technology and Assessment). 1987. Identifying and Regulating Carcinogens. OTA-BP-FI-42-P. Washington, DC: U.S. Government Printing Office.

Renwick, AG. 2000. The use of safety or uncertainty factors in the setting of acute reference doses. Food Addit. Contam. 17(7):627-635.

Renwick, A.G., and N.R. Lazarus. 1998. Human variability and noncancer risk assessment—an analysis of the default uncertainty factor. Regul. Toxicol. Pharmacol. 27(1 Pt. 1):3–20.

UN (United Nations). 2003. Globally Harmonized System of Classification and Labelling of Chemicals. United Nations Economic Commission for Europe [online]. Available: http://www.unece.org/trans/danger/publi/ghs/ghs_rev00/00files_e.html [accessed April 1, 2005].

6

New Approaches

The committee's review of current toxicity-testing strategies revealed a system that is approaching a turning point. Most toxicity-testing frameworks were developed decades ago and may not adequately reflect today's science, let alone the emerging challenges and new approaches of the future. Agencies have responded by altering individual tests and by adding tests to the existing regimens to incorporate end points and mechanistic evaluations newly recognized to be of potential importance. Those patches have not provided a fully satisfactory solution of the fundamental problem.

The core of the problem appears to be tension among four objectives of regulatory testing schemes that are difficult to meet simultaneously: depth, providing the most accurate, relevant information possible for hazard identification and dose-response assessment; breadth, providing data on the broadest possible universe of chemicals, end points, and life stages; animal welfare, causing the least animal suffering possible and using the fewest possible animals; and conservation, minimizing the expenditure of money and time on testing and regulatory review (see Figure 6-1). The committee initially noted that decreasing animal use and decreasing costs may pull the testing programs in similar directions, such as toward the use of in vitro methods. However, those two objectives may not always be aligned; initial efforts to reduce animal suffering and animal use may increase costs in some situations. Thus, approaches designed to move toxicity testing toward one of the objectives frequently move it away from one or more of the others. The Environmental Protection Agency (EPA) and other agencies that perform or require toxicity

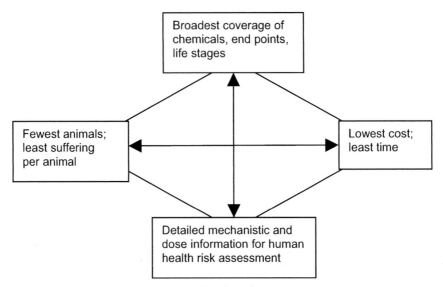

FIGURE 6-1 The four objectives of toxicity testing.

testing are constantly being challenged to meet all four objectives and are often caught between competing priorities.

Setting priorities among the competing objectives is more than a scientific issue. Individuals and organizations can have different, sometimes strongly held beliefs about which of the four is most important, and trying to satisfy the objectives has driven different efforts to reform toxicity testing. For example, legislation requiring creation of an endocrine-disruptor screening program was driven by an effort to increase breadth, whereas the Interagency Coordinating Committee on the Validation of Alternative Methods (ICCVAM) was created in an effort to conserve animals. Rather than attempting to set priorities among the objectives, this committee recognized that all four are important. The committee acknowledges, however, that embracing all four poses a difficult challenge. This chapter and the next review some selected approaches that may ultimately help to move toxicity testing toward one or more of the objectives.

In this chapter, the committee summarizes and comments on some strategies proposed by others for near-term improvements in existing toxicity-testing approaches, including the EPA review of toxicity data available for establishing reference doses (RfDs) and reference concentrations (RfCs), the International Life Sciences Institute (ILSI) Health and Environmental Sciences Institute (HESI) draft reports proposing modifications of EPA's approach for pesticide testing, the National Toxicology Program (NTP) Roadmap for the Future, and the European Union (EU) strategy in the proposed REACH (Registration, Evaluation and Authorisation of Chemicals) testing scheme for existing industrial chemicals.

APPROACHES FOR IMPROVING EXISTING TOXICITY-TESTING STRATEGIES

Toxicity-testing guidelines and strategies described in the previous chapters are the results of the gradual evolution of testing requirements and risk-assessment approaches that took place as the field of toxicology advanced. However, agencies have struggled to incorporate recent scientific and technologic advances in toxicology, basic human biology, molecular biology, pharmacokinetics, dose-response modeling, imaging, computation, and other relevant fields, so many of the current requirements are still based on approaches that originated more than 40 years ago. In addition, more sophisticated exposure assessments have identified different durations and routes of exposure for various populations, such as residential exposures of toddlers, that require more toxicology data for risk assessment.

Proposed strategies for improving toxicity testing can be difficult to compare directly. Some strategies aim at meeting regulatory mandates and therefore focus on specific needs. For example, the ILSI-HESI documents described below focus on toxicity-testing strategies for pesticides, whereas the REACH program attempts to address the numerous industrial chemicals that have been inadequately studied. The different purposes of those testing strategies contribute to major differences between them. However, there can also be important differences between testing strategies and approaches of initiatives and proposals that try to fulfill the same risk-management needs.

The committee elected to focus primarily on the major aspects of the reports reviewed, rather than critiquing the details. Most of the reviewed reports describe initiatives or proposals that are still under development, some of which are sometimes presented with few details; some reports were available to the committee only as drafts. The committee reviewed the documents primarily in an effort to compare various testing strategies proposed and to evaluate their potential to move toward the overall objectives of broadening coverage, increasing depth, addressing animal-welfare issues, and conserving resources. In general, the committee did not engage in detailed assessments of individual bioassays, although a few observations related to the alternatives presented by EPA for discussion are included.

Environmental Protection Agency Review
of Data Needs for Risk Assessment

A Review of the Reference Dose and Reference Concentration Processes (EPA 2002), written by the EPA Risk Assessment Forum's RfD/RfC Technical Panel, provides recommendations for improvements in deriving RfDs and RfCs (see Chapter 5 of the present report for a discussion of RfD and RfC derivation). The committee focused on aspects of the EPA document related to toxicity testing, particularly its Chapter 3, which reviews the adequacy of tests in EPA guidelines (EPA 2005) for deriving RfDs and RfCs for chronic, acute, and other less-than-lifetime exposures.

EPA reviewed information generated from currently required acute, short-term, chronic, and specialized toxicity studies. It then identified data gaps with regard to the assessment of life stages, end points, route and duration of exposure, and latency of response and made recommendations to fill the gaps. Options for alternative testing systems were also presented. EPA (2002) was careful to point out that

> the intent of this review is not to suggest that additional testing be conducted for each and every chemical in order to fill in the information gaps identified for those organ systems evaluated. Nor is it suggested that the alternative testing protocols discussed in this chapter be conducted for every chemical or become part of current toxicology testing requirements or that these alternative protocols are the only options available. Rather, it is the goal of this document to provide a basis for the development of innovative alternative testing approaches and the use of such data in risk assessment.

Specifically, EPA made a number of observations about the generation of toxicity data under the test guidelines for exposures at different life stages. Most of the standard adult toxicity-testing guidelines for acute, subchronic, chronic, and carcinogenicity testing were not designed to evaluate different life stages, and that has led to substantial gaps (see Figure 6-2). Acute and short-term testing is done only in prenatally exposed animals and in young adults, not in postweaning young animals or aged animals. In addition, only a few toxicologic end points are evaluated in the EPA-required acute toxicity (lethality) studies. EPA discusses how data typically collected in subchronic studies—such as hematologic, clinical, and histologic data—could augment acute studies so that acute RfDs could be developed on the basis of end points other than lethality.

Subchronic and chronic toxicity studies are conducted in young adult animals, and exposure in the rodent chronic–carcinogenicity studies continues to the age of 2 years, considered by EPA to be into old age.

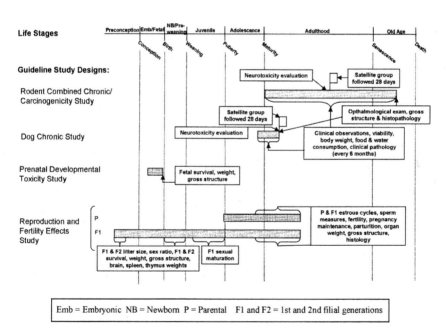

FIGURE 6-2 Guideline study designs used to derive the oral reference dose. Life stages during which exposure occurs (gray), times at which observations are made, and end points evaluated are indicated. Source: EPA 2002.

No subchronic or chronic toxicity studies include exposures beginning in the prenatal period, in the early postnatal period (before weaning), or in animals younger than 6-8 weeks old. Tests for reproductive effects provide data on subchronic exposures of animals that are exposed from before birth up to mating of the F1 males and females and through pregnancy of the F1 young adult females. No subchronic-toxicity evaluations are conducted in aged animals.

EPA emphasized the need to collect pharmacokinetic data that would help define the internal dose of the active agent to the target site. That information could be used to improve study design, study interpretation, dose scaling, and route extrapolation. EPA also notes that there are no guideline protocols for pharmacokinetic evaluations related to exposures and outcomes during infant development or later in old age. The pharmacokinetic and pharmacodynamic datasets are described as useful in determining the interspecies and intraspecies uncertainty factors and in calculating human equivalent exposure concentrations and doses.

As a result of its review, EPA identified several important gaps in the current toxicity-testing framework and concluded that there was minimal evaluation of the following: aged animals in general, but especially after early exposures; some systems and end points (for example, cardiovascular and immunologic) in terms of both structure and function; latency and reversibility; effects of acute and short-term exposure, which are needed to determine acute and short-term RfDs and RfCs; pharmacokinetics; and portal-of-entry effects, especially for substantial dermal exposure.

EPA's recommendations address two main objectives for testing strategies—evaluating a broader array of end points and life stages and increasing information on mechanism or mode of action to improve the human relevance of risk assessment—and include the following:

- Develop a strategy for alternative approaches to toxicity testing, with guidance on how and when to use existing and newly recommended guidelines.
- Develop guidelines or guideline study protocols that will provide more systematic information on pharmacokinetics and pharmacodynamics (that is, mechanism or mode of action), which includes information at different life stages.
- Develop protocols for acute and short-term studies that provide more comprehensive data for setting reference values.

- Modify existing guideline study protocols to provide more comprehensive coverage of life stages for both exposure and outcomes.
- Collect more information on less-than-lifetime exposures to evaluate latency and reversibility of effect.
- Develop guidelines or guideline study protocols to assess immunotoxicity, carcinogenicity, and cardiovascular toxicity at different life stages.
- Explore the feasibility of setting dermal reference values for direct toxicity, including sensitization, at the portal of entry.

EPA explored, but did not endorse, different testing protocols for acute, subchronic, and chronic toxicity testing to address gaps pertaining to life stage, duration of exposure, and latency. Specifically, EPA described an alternative acute-toxicity testing protocol and two alternative chronic toxicity-testing protocols. The purpose of the alternative acute-toxicity testing protocol is to provide hazard and dose-response information after a single acute exposure. That protocol uses a control group and at least three dose groups with 10 animals per sex per group. Clinical signs of toxicity are recorded daily, and food consumption and body weights are recorded on days 1-4, 8, and 14. Five animals per sex per group are killed 3 days after dosing, and the remaining animals are killed 2 weeks after dosing. At both times, urinalysis and hematologic and clinical-chemistry analyses are conducted to address potential reversibility and latency of effects within 14 days of dosing. In addition, the animals are necropsied, organ weights are recorded, and the organs are examined histologically. On the basis of other toxicologic data, this study may be conducted with animals at different life stages and include other end points.

EPA described two chronic protocols with continuous exposure through all life stages. The first is essentially the in utero carcinogenicity evaluation used by the Food and Drug Administration (FDA) for food additives but with yearly interim kills and study termination when the animals reach the age of 3 years rather than 2 years. It involves exposure before mating and then continuous exposure of offspring. In addition to routine clinical pathology and histopathologic evaluations, unspecified neurotoxicity and immunotoxicity testing would be conducted yearly. The second protocol is a unified screening study that has at least four segments: a two-generation reproduction and fertility study, an expanded chronic–carcinogenicity study (Figure 6-3), a developmental-toxicity

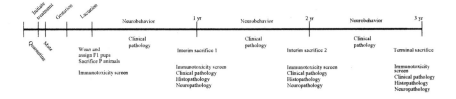

FIGURE 6-3 Expanded chronic-carcinogenicity study. Source: EPA 2002.

study, a developmental-neurotoxicity study, and an optional continuous-breeding study.

EPA identified important challenges to the chronic protocols, including the total number of animals needed and the feasibility of conducting studies as complex and large as the two chronic studies without increasing experimental error. EPA presented the chronic protocols "to demonstrate the advantages (and disadvantages) of exploring nontraditional testing paradigms" and noted that using them in a regulatory setting would require thorough discussion between EPA and registrant scientists.

Committee's Evaluation of Environmental Protection Agency Review of Data Needs for Risk Assessment

Overall, the committee agrees with EPA's analysis and conclusions regarding data gaps. For the most part, EPA's recommendations, if implemented, would improve RfD and RfC development. They encourage the development of innovative toxicity-testing protocols, and following EPA's recommendations for any one chemical would enhance the depth of information on pharmacokinetics, life stages, and end points available for hazard identification and dose-response assessment. However, such an intensive toxicity-testing approach would probably be applied to only a small fraction of chemicals in commerce, would increase the number of animals used to study one chemical, and might even reduce the numbers of chemicals tested. The EPA analysis and recommendations do not address the overall goals of increasing the breadth of coverage of chemicals, conserving animals, and reducing costs and other expenditures. Ultimately, the EPA recommendations need to be evaluated within a larger strategy that considers the uses of toxicity testing for purposes other than to support quantitative risk assessments (for example, chemical-class testing to guide broader screening and coverage).

EPA also made recommendations for research on uncertainty factors used in dose-response assessment. The committee agrees with EPA's conclusion that it is important to research the basis of uncertainty factors. For example, research on the intraspecies uncertainty factor is needed to evaluate whether it adequately addresses the full range of variability of response due to different life stages, genetic susceptibility, and other factors. However, the committee is reserving its detailed comments on the issues surrounding uncertainty factors for its second report.

Specifically, the committee identified five major issues raised in the EPA review that require evaluation and comment: (1) the presence of data gaps in current toxicity-testing approaches, (2) a possible need to refine acute-toxicity testing protocols to support short-term risk assessments, (3) concerns about methods to incorporate pharmacokinetic and pharmacodynamic data into toxicity-testing approaches, (4) questions regarding incorporation of data on direct dermal toxicity into RfD development, and (5) a need to reconsider current toxicity-testing strategies systematically with an eye to improving efficiency and effectiveness. In addition to comment on those major issues, the committee offers some general observations on the acute and chronic toxicity-testing protocols that are explored in Section 3.3 of the EPA report as alternatives to guideline studies.

Presence of Data Gaps

The committee agrees with EPA that there are numerous data gaps in life stages and end points covered in current testing approaches and in functional assessments of some organ systems. However, there are insufficient data to determine the degree to which those data gaps have practical significance in risk assessment or whether they are primarily of theoretical or academic concern. Depending on how the data are used in risk assessment, some of the data gaps may have little effect on the final outcome, whereas others may be very important. The committee cautions against adding testing requirements only for the sake of thoroughness, because such an approach can result in a substantial waste of animals and resources with little gain. The challenge is to try to cover a broader range of end points and life stages with improved sensitivity to subtle or functional effects without indiscriminately expanding current toxicity-testing approaches.

The committee favors a two-pronged approach in the near term. Such an approach includes relatively modest enhancements of current protocols designed to improve screening for a somewhat wider array of end points in a somewhat broader array of life stages. Such enhanced protocols could trigger more in-depth testing. Adequate second-tier evaluation of such end points as immunotoxicity and cardiovascular toxicity may require the creation of new test guidelines with indications of what would trigger such testing. Research is needed to determine the practical effect of filling or not filling some of the data gaps. In the meantime, it is presumed that current uncertainty and adjustment factors applied by EPA to derive reference values are sufficiently large, although that conclusion has not yet been established. The magnitude of uncertainty factors used in deriving reference values can be reassessed once the enhanced protocols are developed, implemented, and validated and a comprehensive review of outcomes is conducted for a set of test chemicals.

It is also important to look beyond animal toxicity data for help in resolving some of the questions. Epidemiologic studies with reliable exposure assessment can shed some light on the likelihood that current toxicity-testing data are missing important end points or are insufficiently sensitive to be applied to life stages not studied. For example, the prospective National Children's Study[1] or other prospective cohort studies that include children and the elderly may be helpful in ascertaining differences in susceptibility. Studies of workers who handle various agents can be of great help in assessing effects on the immune system, cardiovascular system, and functional end points that are not now well assessed with animal studies. However, human data will be available on only chemicals that are already in use and, because of problems with exposure assessment and other difficulties in epidemiologic studies, can be collected only on a small subset of existing chemicals. In addition, legal and ethical issues limit the design of human studies that can be conducted and integrated into toxicity-testing strategies.

---

[1]The National Children's Study can be found at http://nationalchildrensstudy. gov/.

Refinement of Acute-Toxicity Testing Protocols

The committee agrees with EPA that the current protocols for acute toxicity testing focus on lethal effects and gross observations and generally do not provide adequate information for acute and short-term RfDs and RfCs. EPA has identified a regulatory need for improved data for setting acute RfDs and RfCs, and such data are within reach of current toxicity-testing approaches. Improved acute-toxicity assessment may be particularly important in situations where there is high worker exposure and for the development of acute-exposure guideline levels (AEGLs) used to design emergency response plans for accidental releases of chemicals. However, conducting acute protocols that address latency, reversibility, and differential susceptibility for all toxicity end points currently required in subchronic and chronic protocols (such as hematology, clinical chemistry, pathology, functional tests, and detailed clinical observations) will lead to very complex animal studies. Before such complex protocols are conducted, acute LD_{50} studies, repeated-dose toxicity studies, and human data should be evaluated to determine the need for these studies and ultimately to guide the design of such studies.

Pharmacokinetics[2] and Pharmacodynamics

The committee agrees with EPA that generally little information is available on pharmacokinetics, including possible differences across life stages. Although there is a need to develop more information on pharmacokinetics and pharmacodynamics, it is critical to define the purpose of such studies to avoid the creation of data that are unlikely to be used and that constitute a waste of animals, time, and resources. Data on absorption, distribution, metabolism, and excretion (ADME) should be used in guiding toxicity testing, including identifying the most relevant routes of exposure. Beyond the ADME studies, additional data should not be required routinely but instead case by case. It is important to use caution in drawing major conclusions on the basis of relatively few pharmacokinetic and pharmacodynamic data. For example, such data may not always be sufficiently explanatory to support major conclusions about

---

[2]*Pharmacokinetics* as used in this chapter encompasses both the quantitative and qualitative aspects of absorption, distribution, biotransformation, and excretion of chemicals.

concordance or extrapolations to other routes of exposure. It is often tempting to assume that one mode of action that has a strong hypothetical basis supported by some data fully explains the toxicity of a given chemical. However, many chemicals have multiple modes of action, and these can depend on exposure route, concentration, and duration. Furthermore, there can be competing plausible hypotheses that may not have been considered or subjected to study.

Dermal Portal of Entry

The committee considered EPA's conclusion that there is a need to improve the assessment of portal-of-entry effects, especially on the skin, in the risk-assessment process. EPA emphasized dermal effects because oral and inhalation effects are already recognized as important routes of exposure and require specific test data for risk assessments in, for example, its pesticide testing programs. Current protocols for dermal exposure, when they are applied, do evaluate dermal toxicity, including sensitization, but the overall assessment of systemic and portal-of-entry effects after dermal exposure is limited. The committee agrees with EPA that there is a general need for better dermal-uptake data and for a better understanding of uptake after injury to the skin. However, it is important to consider whether skin is an important route of exposure before triggering an effort to set a dermal RfD. Again, human data can be useful in assessing the effects of dermal exposure. Worker data and clinical reports have been among the main sources for identifying dermal irritants and sensitizers. Such data could be collected more systematically and used preferentially in setting dermal RfDs for existing chemicals. In addition, there should be a more systematic effort to gather postmarketing data on chemicals to ascertain, for example, whether dermal toxicity is reported.

Toxicity-Testing Strategy

The first recommendation of Chapter 3 of the EPA review is that EPA "develop a strategy for alternative approaches to toxicity testing, with guidance on how and when to use existing and newly recommended guidelines" (EPA 2002). The committee agrees that such a new strategy is needed to improve efficiency, reduce animal use, increase the number of chemicals that are screened for toxicity, and address some of the data gaps noted above. Such an improved strategy could include setting priorities for screening and testing of agents and incorporating such factors as use, exposure, pharmacokinetic data, and new screening tools. The committee's second report will be designed to help to create just such a new strategy.

Specific Testing Protocols

EPA explored alternative testing protocols for acute and chronic toxicity testing. In making the following comments on those protocols, the committee recognizes that EPA offered them primarily to stimulate new ideas. In addition, the committee recognizes the difficulty of evaluating specific protocols out of the context of the overall testing strategy.

Acute-Toxicity Testing Protocol

EPA proposes an alternative acute-toxicity testing protocol that provides more-comprehensive toxicity data for derivation of the acute RfD than its current acute-toxicity test guidelines (EPA 1998a,b,c) that provide data to calculate acute $LD_{50}s$ and $LC_{50}s$ and to select doses for use in other studies. Taken in isolation, the alternative toxicity study may not fully address all the important data gaps raised by EPA in its analysis of acute toxicity tests, such as reversibility, latency, and end-point assessment in detail and at different life stages. The animal group size of five is small for conducting hematology, clinical chemistry, and pathology evaluations, and this limits the capacity to detect effects, particularly subtle ones. For example, statistical significance of dichotomous findings of effects in a group of 10 animals requires that 40% of the animals be affected if the effect rarely occurs in treated animals; greater proportions are required for more common effects or smaller groups. Even so, compared with the current acute toxicity tests, the protocol is relatively resource-intensive with respect to animal use and histopathologic examination.

Designing comprehensive acute toxicity testing could be guided by existing data, including data from repeated-dose studies. For example, acute risk assessment could first be conducted on the basis of short-term repeated-dose studies and developmental-toxicity studies. When human exposures are well understood, the margin of exposure could be examined. If the margin of exposure is too small, sensitive acute toxicity studies could be designed and performed to focus on the more sensitive end points of concern. Indicators of acute toxicity could also be added to repeated-dose studies, such as an early assessment of hematology and clinical-chemistry end points and more frequent clinical observations on the first few days of repeated-dose oral, inhalation, or dermal toxicity studies. The limitations of adding early assessments to repeated-dose studies are that not all acute end points may be identified and the dose-response relationships may not be adequately characterized, especially those which might result from higher concentrations that cannot be used in longer-term repeated-dose toxicity studies.

Alternative Chronic Testing Protocols

EPA also presented two chronic protocols that address continuous exposure through all life stages as alternatives to current test procedures. The chronic–carcinogenicity study would evaluate the potentially increased sensitivity of both developing and aging animals to chronic and carcinogenic effects of chemicals. It would be relevant to chemicals to which there may be substantial chronic exposure of the general population or of workers. Studies of that type have already been conducted by industry and contract laboratories.

According to current combined chronic–carcinogenicity test guidelines (EPA 1998d), 50 rats per sex per dose are required, for a total of 400 animals. The expanded study specifies 25 rats per sex per dose for each annual interim kill, for a minimum of 500 animals. In practice, more animals may be necessary to account for the possibility of decreased survival. EPA addressed the issue of increased animal use, in part, by suggesting fewer animals per segment (for example, 20 per sex per group) or only one interim kill (for example, at 1.5 years). EPA acknowledged that reducing the number of animals to 20 per sex per group decreases the statistical power to detect tumors but did not suggest further study changes to resolve the issue satisfactorily.

Another challenging aspect of the study is setting the doses to produce sufficient toxicity while yielding an adequate survival rate by the age of 3 years. Setting doses too low results in a study that is not sufficiently sensitive to detect an effect; setting doses too high can increase mortality, leaving the group size reduced and thereby compromising the statistical power to detect effects. Also, the feasibility of conducting a study of a specified duration will depend on the survival characteristics of the strain chosen, which can range from less than 2 years to 4 years for different rat strains. Instead of selecting a predetermined termination time, the study could end when some prespecified fraction of control or treated animals is surviving (for example, 20%).

Another limitation of the expanded chronic–carcinogenicity study is that it may not provide an adequate basis for quantifying increased susceptibility to carcinogenicity at different life stages. That information is important for risk evaluations if people are exposed at different doses at a given age (for example, while breast-feeding and while being bottle-fed). For those reasons, alternative study designs should be considered, such as designs that include dose groups that are exposed briefly and followed for long periods. As alluded to by EPA, the study design makes difficult or impossible the assessment of whether toxicity is due to repeated chronic exposures or to shorter-term prenatal exposures.

Although the chronic–carcinogenicity study is a challenging study design, the committee supports the development of bioassays that address differential sensitivity to a chemical's immunologic, neurologic, and cancer effects during different life stages. However, this study is limited in that it may be more difficult to attribute effects to any one life stage (that is, prenatal development period, the postnatal period before weaning, adolescence, and old age) and would not be useful for shorter-term risk assessment for different populations.

EPA presented the Unified Screening Study to stimulate ideas on how studies could be combined to limit animal use while expanding the toxicity evaluation across multiple generations to study all life stages and transgenerational effects. Further development will require considerable resources and expertise in the fields addressed by the different arms of the study. It would require development and pilot experimentation by such a group as the NTP before such a proposal could be implemented. The committee's review below highlights a few important issues that may arise from conducting such a study.

The Unified Screening Study attempts to economize animal use in a multigeneration study so that fewer naïve animals would be used than if all the studies that would potentially be replaced by it were done separately. Substantially more animals would be needed than are now required in a two-generation study to ensure that sufficient numbers of animals were available for both the chronic study and the reproduction study and to guarantee survival of enough animals to the age of 3 years.

EPA (2002) states that animals "assigned to the chronic/carcinogenicity segment should be genetically diverse within each dose group and should ... not be siblings." For that segment, only one male and one female from each litter could be used to meet the need for 100 males and females per dose group. Thus, at least 100 litters per dose group would be needed. That is 5 times the number of litters required by the standard reproduction study (EPA 1998e) and would make such an experiment difficult to conduct. Further evaluation is needed regarding the importance of the requirement of genetic diversity in the cancer-bioassay segment of an expanded study like this one.

As with the expanded chronic–carcinogenicity study described above, a major challenge is dose selection. It is compounded by the fact that there are many arms to the study. Because many more animals per litter would be assigned to either the chronic study or the reproduction study, there is less room for error in setting doses. EPA (2002) states that "although it is assumed that treatment levels and route of administration will remain constant across all study segments, this approach to dose-setting and route selection may not always be optimal for every phase." That may be especially true for dietary exposures in which chemical exposure of pups around the time of weaning can be much higher than exposure of adults and result in unacceptable toxicity in pups. Insufficient animals would then be available for both the two-generation reproduction study and the chronic–carcinogenicity study. Selection of dose is difficult for each of the studies when considered separately and will be exceedingly difficult for such a complex study. Setting doses either too high or too low could lead to repeat studies or addition of dose groups. Keeping doses the same in the different arms would generally mean lowering them to preclude toxicity in the most sensitive study arms, and that would reduce study sensitivity overall and decrease one's ability to characterize the dose-response relationship.

As noted by EPA, another challenge in the development of the Unified Screening Study is the selection of strain. Sprague-Dawley and Wistar rats (outbred strains) are typically used in toxicity studies, including those of developmental toxicity, reproductive toxicity, and developmental neurotoxicity. Inbred Fischer 344 rats are often used in standard chronic–carcinogenicity studies. Thus, the Unified Screening Study would at a minimum require a change in the standard strain used (that is, for either carcinogenicity or reproductive and developmental-toxicity end points). EPA noted that changing the strain used in any type of study would compromise the use of historical data. Furthermore, a sensitive strain to detect carcinogenicity (NTP 1984, 2005) may not be the optimal strain for developmental and reproduction studies or the optimal strain for longevity. The NTP is reviewing the issue of strains used in chronic testing, and results of this effort should be useful to EPA as it considers alternative test protocols (NTP 2005).

The idea of obtaining animals for the developmental-neurotoxicity segment from the reproduction and fertility study segment's second generation (F2) is more complex than is presented by EPA. The major difficulty in the developmental-neurotoxicity segment is the need to maintain tight control of mating to manage logistically the multiple behavioral testing required during lactation and immediately after weaning. Lack of tight control over mating can easily cause behavioral testing for one time (for example, postnatal day 13) to spread out over many days and overlap into other testing periods, resulting in increased variability of the data. Such a study can be conducted, but it is not optimal for developmental neurotoxicity.

A protocol of the magnitude and complexity of the Unified Screening Study would require substantial development and evaluation before it could be implemented as a test requirement. EPA does not indicate exactly how the protocol would be applied. Some committee members found that routine use of the study as a first-tier screening test would be problematic for both scientific and practical reasons. They believed that potential disadvantages of the study outweigh its advantages, especially because it appears to preclude the use of toxicity results from one study to trigger or prevent additional evaluation in other studies in that all studies would be conducted simultaneously. Other members of the committee found that the study protocol could function as a component in a toxicity-testing strategy as applied to either selected chemicals with widespread human exposure or chemicals with initial indications of developmental toxicity. Some segments of the study could be applied in a tiered fashion, in which selected end points could be examined in followup studies if triggered by observations in earlier segments.

In addition to the scientific and strategic concerns, there are practical concerns that the studies are extremely resource-intensive and complex. The separate components are temporally tied to each other intimately, thereby decreasing flexibility of study starts in any laboratory that might be capable of conducting all arms of the experiment. Furthermore, few laboratories have the capability (for example, trained personnel and appropriate computer systems) to conduct multigeneration, developmental-neurotoxicity, and immunotoxicity evaluations. Hence, widespread replacement of current approaches by the Unified Screening Study could reduce the number of chemicals evaluated during the development of laboratory capacity.

In conclusion, EPA has explored alternative testing protocols but does not articulate how they would be incorporated into a testing strategy. The committee supports the notion of expanded tests that combine studies to limit animal use and provide more in-depth evaluations of end points and life stages. However, considerable development and evaluation may be required to ensure that tests are feasible, do not compromise study sensitivity, produce the desired data, and reduce the use of animals. The committee also points out that subchronic and chronic studies might be expanded to obtain better data for establishing acute reference values. The expanded bioassays may ultimately have a role as part of a broader testing strategy applied selectively to high-priority chemicals, but such approaches may not be amenable to widespread application.

International Life Sciences Institute Draft Proposals[3]

ILSI-HESI developed three draft white papers (ILSI-HESI 2004a,b,c) proposing a tiered-testing scheme that includes several tests that are currently part of EPA's testing scheme for pesticides but also deletes some that are required. Modified protocols for some EPA guideline toxicity tests are also included. ILSI-HESI focused on EPA toxicity-testing approaches for pesticides, but in principle the approaches could apply to other compounds that are subject to toxicity testing.

One ILSI-HESI paper focuses on assessing systemic toxicity in young adult animals. For the purposes of the project, ILSI-HESI (2004a) defines *systemic toxicity* as "the potential adverse effects of agricultural chemicals on 'young adults.' " The evaluation of effects of chemicals on different life stages—including reproduction, development, adolescence, and old age—is presented in a second paper. The third paper focuses on the acquisition and application of pharmacokinetic data in agricultural-chemical safety assessments. The overall testing scheme incorporating the three aspects is outlined in the second draft paper, on life stages.

---

[3]A number of details in the ILSI-HESI draft white papers require careful scrutiny before implementation (for example, reduction of premating period to less than half the spermatogenic cycle in the extended one-generation study). The committee did not review proposed study protocols in detail, and its lack of comment should not be taken as agreement.

The discussion of the testing scheme below focuses on the general concepts and approaches rather than the technical details of the draft papers.

Systemic-Toxicity Testing

ILSI-HESI proposes a tiered approach to systemic-toxicity testing that considers the use profile, the resulting exposure, and the toxicologic profile of the pesticide in identifying test requirements (ILSI-HESI 2004a). The goal is to improve the adequacy of the dataset for the variety of risk assessments required while reducing test artifacts and the number of animals used in testing and increasing the efficiency and accuracy of the safety-assessment process. The overall process used by ILSI-HESI included (1) identifying exposure durations that are important for pesticide risk assessment, (2) evaluating a database of EPA guideline test results on representative pesticides to determine the necessity of particular tests by evaluating how EPA used test results in setting the RfD, (3) developing or modifying toxicity tests to address risk-assessment needs, and (4) developing a flexible tiered-testing scheme that would be guided by pharmacokinetic data and the known toxicologic profile of the chemical.

Originally, the assessment of agricultural chemicals focused on continuous dietary exposure; that led to an emphasis on lifetime exposures and acute accidental exposures. Today, EPA conducts risk assessments to evaluate a variety of agricultural and residential pesticide exposures of different durations and at different life stages. ILSI-HESI identified the following human exposure durations for which risk assessments may be required: 1 day, 2-28 days, 1-6 months, greater than 60 months, and intermittent. ILSI-HESI compared those durations with current toxicity-testing requirements and concluded that the tests did not meet all the needs of risk assessors.

ILSI-HESI evaluated the toxicology database of results of subchronic and chronic pesticide studies to examine which studies were used most often in deriving the RfD or in classifying carcinogenicity. The analysis was used to differentiate between study types that should be retained or developed further for future use and types that should not. Of the hundreds of pesticides in the EPA database, data on 65 pesticides representing different classes were entered into a toxicity database developed specifically for the ILSI-HESI evaluation. Of the 65 pesticides selected, data on 28 were sufficiently robust to allow the kinds of comparisons that ILSI-HESI intended to make. Results of subchronic and chronic studies were compared to examine the relative sensitivity of different test species and different exposure durations: 90-day studies in rats versus dogs, 90-day versus 1-year studies in dogs, 90-day versus 2-year studies in rats, 2-year rat versus 1-year dog studies, and rat versus dog studies as the basis of the chronic RfD. The effect of mouse carcinogenicity studies on cancer-hazard determinations, but not dose-response assessment, was also assessed.

For the dataset of 28 chemicals, the chronic RfD was based half the time on the rat and the other half on the dog. Significant quantitative differences were noted; most often, the dog was more sensitive in the 90-day study. It was concluded that both species should be used in pesticide testing. In the rat, the no-observed-adverse-effect levels (NOAELs) in 2-year studies were often lower (by more than a factor of 2) than those in the 90-day studies; in the dog, that was also the case for the 1-year versus 90-day study but somewhat less in frequency and magnitude. On the basis of those data, ILSI-HESI concluded that the 1-year chronic dog study did not affect the RfD assessment in an important way. Finally, the mouse carcinogenicity study was proposed for elimination because it was found to add little to the findings of the rat carcinogenicity study.[4]

After considering possible test redundancies and risk-assessment needs, ILSI-HESI selected core tests for initial toxicity evaluation. For assessing short-term exposures, a new 28-day rat study with recovery period and a modified 90-day dog study were proposed. ILSI-HESI proposed that single-exposure acute studies be performed if the NOAELs from those studies were too close to the predicted single-day human exposures.

The results of the 90-day dog and 28-day rat studies are compared, and the more sensitive species is taken as the more relevant unless compelling scientific data indicate otherwise. If the rat is considered the relevant species, the new 28-day rat study is used to assess the effects of human exposures of less than 6 months, and a 1-year rat study for exposures of more than 6 months. The 1-year rat study would be conducted as part of an expanded 2-year carcinogenicity assay; an interim kill would occur at 1 year. If the dog is found more relevant, the 90-day dog study would be used to evaluate effect of human exposures over 2 days

---

[4]One important inconsistency between the draft papers is that the life-stages paper appears to eliminate carcinogenicity testing, but the systemic-toxicity paper includes the rat carcinogenicity test in the first tier (see Figures 6-4 and 6-5). That may have been an oversight in the draft life-stages paper. (Since completion of this interim report, the life-stages figure in the final paper [Cooper et al. 2006] was revised and now includes carcinogenicity testing in the first tier.)

and up to 6 months; for exposures beyond 6 months, the 1-year rat or 90-day dog study would be used. A 24-month carcinogenicity study would also be available. The 28-day rat or 90-day dog studies would be considered the most relevant for human effects of exposures of less than 6 months. Performance of the longer-term or acute studies would be contingent on predicted human exposure scenarios. Figure 6-4 shows the tiered-testing strategy proposed by ILSI-HESI.

The pivotal studies of the testing strategy are the 28-day rat study and the 90-day dog study. In addition to a 14-day recovery group, the rat study would include satellite groups for neurotoxicity and immunotoxicity evaluations as indicators of specific effects to trigger additional specialized studies. If the indicators were affected in that study, second-tier specialized testing would be performed to investigate indicated end points further or to characterize mode of action. Tissues would be examined as specified in the EPA guidelines for the oral 90-day rat study (EPA 1998f). The 90-day dog study would extend the existing guideline study to include cardiac and pulmonary evaluations and would include

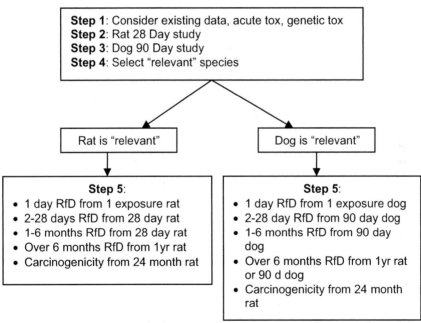

FIGURE 6-4 ILSI-HESI draft proposed tiered approach for systemic-toxicity evaluation. Source: ILSI-HESI 2004a. Reprinted with permission. This figure appears in the final paper of Doe et al. (2006) in *Critical Reviews in Toxicology*; copyright 2006, Taylor & Francis Group, LLC.

blood draws for pharmacokinetic analyses. Treatment-group size would be increased in the dog study from four to six animals.

The ILSI-HESI tiered-testing strategy assumes that genotoxic chemicals would not be developed as pesticides and emphasizes the use of pharmacokinetic data from basic ADME studies to support the design of toxicity studies (for example, using those data to determine the highest dose to test and the duration of the recovery period). ILSI-HESI also acknowledges the potential future use of novel high-throughput screens that can predict potential adverse effects of concern and thereby indicate relevant end points to incorporate into the tier I studies. The new technologies would be derived from in vitro screening, computational biology, molecular biology, and systems biology. ILSI-HESI envisions that high-throughput screening could potentially include in vivo and in vitro biologic systems in which the active ingredient is tested, RNA or protein isolated from the cell culture or target tissues is analyzed, and genomic or proteomic assays are used to identify markers of toxicity that suggest the potential for particular adverse health effects.

Life Stages

ILSI-HESI proposes a tiered-testing approach for "assessing the potential adverse effects of agricultural chemicals on preconception, embryo/fetal and newborn/pre-weaning life stages and on adults of all ages, including the young and aged" (ILSI-HESI 2004b).

The centerpiece of the proposed life-stages paradigm (Figure 6-5) is a new testing protocol for an F1 extended one-generation study. In that study, mature parental males are treated for 4 weeks and parental females for 2 weeks before mating. Treatment is continued for males and females during mating, for females during gestation and lactation, and for males until it is confirmed as not needed for a second mating. Selected F1 pups (three males and three females per litter) are treated continuously until postnatal day 70 and are given a much more comprehensive histopathologic evaluation at postnatal days 21 and 70 than is now required in the multigeneration study (EPA 1998e). Developmental-neurotoxicity assessments (motor activity, functional observational battery, and neuropathology), developmental-immunotoxicity assessments (antibody plaque-forming cell response), and additional toxicity assessments based on the results of a 28-day systemic-toxicity test also would be conducted.

The proposed life-stages tier 1 would include a rabbit developmental-toxicity study and the extended one-generation rat reproduction study.

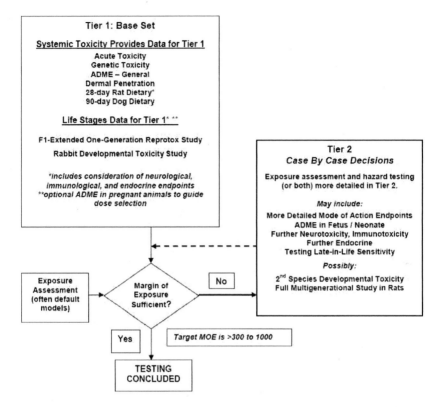

FIGURE 6-5 ILSI-HESI draft proposed tiered life-stages testing scheme for pesticides. Source: ILSI-HESI 2004b. Reprinted with permission. This figure has been modified and appears in the final paper of Cooper et al. (2006) in *Critical Reviews in Toxicology*; copyright 2006, Taylor & Francis Group, LLC.

The first tier incorporates a number of features from the current EPA reproduction and developmental test guideline studies, such as evaluation of comprehensive reproductive and developmental end points. It also includes end points and evaluations not required by EPA guidelines, such as more comprehensive postnatal histopathologic evaluation; neurologic, immunologic, and endocrine end points after exposure during gestation and lactation; and pharmacokinetic observations at key stages of gestation and lactation.

EPA now requires a two-generation rat reproduction study and developmental-toxicity studies in two species. The ILSI-HESI draft paper discusses the potential concerns regarding removal of the second-generation part of the multigeneration study and the rat developmental-toxicity study. ILSI-HESI concluded, on the basis of an unpublished EPA evaluation of the existing pesticide database, that the second generation had relatively minor effects on risk assessment. It hypothesized that rat developmental toxicity may be adequately addressed with the extended one-generation rat study. Additional developmental-toxicity or multigeneration studies in the second tier could be triggered if the margin of exposure were not sufficiently large (for example, less than 300), if there were a substantial difference between the rat and rabbit NOAELs, or if there were effects in rabbit fetuses of a type unlikely to be detected postnatally in rat pups. In addition, extension of the one-generation study into a multigeneration study would be triggered on the basis of potential concerns about possible effects on reproduction, such as alterations in anogenital distance, timings of vaginal opening or preputial separation, or altered estrous cycles.

The ILSI-HESI approach was designed to reduce the number of animals used, the cost, and the time involved in testing while improving the data on several specific end points available for risk assessment. The proposed tiered approach also incorporated an evaluation of margin of exposure with hazard and dose-response assessment to determine whether additional testing would be needed. Finally, the ILSI-HESI draft paper concluded that the intraspecies uncertainty factor of 10 was sufficient to address risks to the elderly on the basis of few toxicity data on aged rats and human data. ILSI-HESI argued that pharmacokinetic differences due primarily to reduced renal function are what drive susceptibility. Relevant triggers for further unspecified animal studies in aging animals include effects on renal function, the immune system, the nervous system, the cardiovascular system, and the liver. The proposal did not specify how and when additional tests would be triggered.

Pharmacokinetics

The ILSI-HESI report *The Acquisition and Application of Absorption, Distribution, Metabolism and Excretion (ADME) Data in Agricultural Chemical Safety Assessment* contains recommendations for generating and using pharmacokinetic information in risk assessment (ILSI-HESI 2004c). It suggests a tiered approach that consists of basic, intermediate, and advanced studies: basic studies to support dose selection and evaluation of half-life; intermediate studies to guide study interpretation and to provide information on absorbed dose, tissue persistence, and tissue accumulation; and advanced studies to aid in understanding of mode of action and to support extrapolations across routes of exposure and among species.

ILSI-HESI argues that it is important to base the dose, route, and species used in toxicity testing on an understanding of a compound's behavior—such as saturation of absorption, elimination pathways, and metabolic profiles—pover a range of doses. The proposed basic tier includes studies to collect blood time-course data after single oral and intravenous exposure and other pharmacokinetic studies to establish the extent and rate of absorption, bioavailability, general distribution of the chemical, extent of metabolism, routes and rates of excretion, half-life, tissue persistence, and clearance. The proposed second tier includes routine metabolite identification and pharmacokinetic sampling during toxicity studies and studies of tissue distribution, serum protein binding, and in vitro metabolism in rodents and other species. If first-tier findings indicate biliary elimination or enterohepatic recirculation, followup studies in the second tier would be conducted. The third tier could include studies aimed at better understanding of route-specific pharmacokinetic differences and further studies to support the application of physiologically based pharmacokinetic (PBPK) models in risk assessment.

Committee's Evaluation of International Life Sciences Institute Draft Proposals

Through a tiered-testing scheme, ILSI-HESI proposes substantive modifications of EPA's toxicity-testing approach for pesticides. In developing its proposal, ILSI-HESI identified some potential omissions and redundancies and possible improvements in existing tests. Recommendations include changing required toxicity-testing durations, removing some currently required guideline studies, modifying other studies to increase end-point coverage, triggering specialized studies on the basis of findings in the core set of toxicity tests, and generating chemical-specific pharmacokinetic data to inform study design and data interpretation. Several recommendations address the goals of conserving animals and other resources and increasing the depth of some particular areas, such as neurotoxicity. The new study designs would result in coverage of a broader array of end points, improved sensitivity to subtle or functional effects, and information on reversibility and latency after 28-day exposures. However, more rigorous evaluation is needed to determine whether the expanded designs compensate for the elimination of particular studies, and some important data gaps with regard to life stages were not adequately discussed.

Several studies that are required in EPA pesticide testing are not included in the first tier of the ILSI-HESI proposal: the rat teratology, 1-year dog, mouse carcinogenicity, and two-generation reproduction and fertility studies. The following discussion provides committee comments on the removal of those studies and highlights the committee's concerns.

As pointed out by ILSI-HESI, no single species has been shown to be the most predictive of human teratogenicity. ILSI-HESI proposes removing the rat teratology study and using the extended one-generation study and rabbit teratology study to evaluate developmental effects. Although the proposed one-generation study substantially improves postnatal evaluation of many nonreproductive end points after in utero and postnatal exposures, it is unclear whether the proposed study would be as sensitive as a rat teratology study for prenatal developmental-toxicity end points or would adequately reveal the potential hazard and trigger a followup study. Furthermore, EPA often bases acute reference values on the rat teratology study. In contrast, postnatal effects in a one-generation study are not typically used for deriving acute reference values. The effect of eliminating the rat teratology study on hazard identification and setting of the acute reference value should be evaluated if the proposal is pursued. Careful validation using known development toxicants is needed.

On the other hand, the proposed extended 28-day rat study, with improvements as discussed above, should provide a better means of assessing short-term exposures than the 90-day rat study and the existing acute toxicity studies used to establish LD_{50}s. The 28-day rat study is also being proposed to assess risks posed by human exposures as long as 6 months. The extent to which the 28-day study improves on the 90-day rat study for exposures longer than 1 month is unclear, and further effort to assess this issue is needed. In a similar vein, the extended 90-day dog study with its improved evaluation of critical end points provides a better subchronic dataset than the current guideline 90-day dog study but does not offer substantial advantages over the current study regarding exposure-duration data gaps. Because the 90-day dog study and the 28-day rat study are proposed to provide core data for assessing human exposures of up to 6 months, further assessment of their use for long-term human exposures is needed; this is critical for the dog study because in cases where the rat is not a good model the 90-day dog study potentially could serve as the basis of the RfD for lifetime human exposure.

The proposal to eliminate the chronic 1-year dog study and the mouse carcinogenicity study is based in part on a review of a dataset on only 28 pesticides. The committee agrees with the overall approach of reviewing the redundancy and importance of tests with retrospective analyses of pesticide-study findings, but it finds that the review in the proposal is too limited to support firm conclusions. For example, for the 21 cases in which comparisons could be made, there were eight cases (38%) in which the NOAEL from the 1-year dog study was less by more than a factor of 2 than that from the 90-day dog study. Although some of the differences may be due to dose spacing in the studies, the differences indicate the ambiguous nature of the comparison and the need for a larger, systematic evaluation. Similarly, the carcinogenicity comparison was based on only 23 chemicals for which rat and mouse carcinogenicity studies were conducted—five showing tumors only in the rat, four only in the mouse (three in liver), six in both species, and eight without tumors—again too few comparisons to support firm conclusions. Quantitative comparisons of dose-response characteristics on the basis of the benchmark dose or cancer potency were not made. Thus, the retrospective review was not sufficiently extensive or rigorous to support study removal.

Nonetheless, the committee strongly supports the general approach used by ILSI-HESI of evaluating existing databases in considering testing strategies, and it recommends a further review based on a larger, more robust dataset. A database of test results from guideline and special studies of the numerous pesticides in the EPA database could be used to evaluate the redundancies and gaps in pesticide testing further. Shorter-term studies, such as 28-day studies (for example, OECD Test Guideline 407), should be evaluated in addition to 90-day and carcinogenicity studies. It would also be informative in assessing study sensitivity to compare benchmark doses in addition to NOAELs and lowest-observed-adverse-effect levels (LOAELs). When comparisons of relative sensitivity across species are made, those evaluations should consider what the most appropriate dose scale (for example, $mg/kg^{3/4}$ per day vs mg/kg per day) should be. Ideally, such evaluations would be conducted blind, without knowledge of the particular pesticide being reviewed. The committee endorses further broad retrospective reviews extending beyond the pesticide database to evaluate study reproducibility and predictability among and between in vivo and in vitro findings. In conducting and assessing retrospective analyses, it should be recognized that the power of human studies is often too low to be of value for evaluations of predictability and study validation. Ultimately, reviews should consider how regulatory decision-making might change with the elimination of some test protocols and the role that test redundancy plays in safety assessments.

The EPA (1998g) approach in its guidelines for pharmacokinetic testing of pesticides is tiered and focuses on the generation of data to aid in interpreting toxicity tests and risk-assessment applications. The ILSI-HESI proposal also is tiered and focuses on generation of similar data, but it emphasizes the generation of pharmacokinetic data to aid in dose selection and other aspects of study design, which is also an important objective. The ILSI-HESI proposed first tier is similar to EPA's first tier, which provides a minimal dataset from standard chemical disposition ADME studies. EPA's second-tier testing is to address specific questions of importance for risk assessment that arise from the toxicology database on the chemical and the tier 1 testing. It can include data generation for PBPK modeling. EPA's second-tier testing requirements are not fixed and can be developed through cooperative agreements with industry. Unlike EPA's, the ILSI-HESI second tier includes routine metabolite identification and pharmacokinetic sampling during some toxicity tests. To ensure study relevance, both EPA and ILSI-HESI schemes emphasize an iterative process of data collection and use of data analysis in designing pharmacokinetic studies for specific compounds; no single prescribed series of studies is likely to provide necessary and adequate pharmacokinetic datasets on every compound.

The possibility of increases in susceptibility to cancer at particular life stages was acknowledged. The ILSI-HESI proposal assumes that genotoxic chemicals will not be developed as pesticides and that the end points added in the one-generation study will provide adequate indication of toxicity and carcinogenicity to trigger necessary testing involving perinatal exposure; however, ILSI-HESI did not identify the indicators that might trigger a perinatal carcinogenicity study. Some nonmutagenic compounds, such as estrogen-receptor agonist and antagonists, may elicit effects in critical windows of susceptibility, and the possibility of tier 2 testing was raised if concerns were triggered by mode-of-action information, but no specific proposals were made. Similarly, no specificity was given regarding when to follow up on possible enhanced sensitivity in aging animals.

The tailoring of the testing regime to meet risk-assessment needs is a strength of the ILSI-HESI proposal. Exposure considerations, such as the margin of exposure between doses that produce effects in animals and expected human exposures to pesticides, also provide a conceptual framework for guiding the selection and extent of testing. The ILSI-HESI proposal to trigger further testing when the margin of exposure falls below some threshold value is a reasonable approach to limiting testing to issues of public-health importance. Still, although using exposure to guide selection of studies may work well for chemicals to which exposure can be relatively well defined, it may not work as well for industrial chemicals, because the degree and circumstances of human exposures can be difficult to predict or adequately assess.

To conclude, the proposed changes in the scheme for pesticide testing may affect the probability of finding some effects and change the volume of evidence available to an assessor in judging the presence or importance of an effect. The changes include eliminating the second species in teratogenicity and carcinogenicity testing, using the 90-day dog study as an indicator of chronic toxicity, and altering dose selection. Cumulatively, it is unclear how the different aspects of the proposal would affect the overall fidelity of the testing process. Thus, another important issue for evaluation is the overall effect of the changes on the sensitivity and reliability of the testing process. The committee notes that redundancy of testing is a critical part of the weight-of-evidence approach that may have been overlooked in the ILSI-HESI evaluation. More-limited testing and less redundancy could mean less confirmatory evidence and greater potential overall for reduced sensitivity of the testing strategy. Making decision-making more conservative, erring in the direction of false positives, or using greater uncertainty factors may address those issues, as would corresponding adjustments of risk-assessment guidance documents that emphasize positive results of multiple studies for confirmatory evidence.

REACH Program

The European Union's REACH program was adopted by the European Commission on October 29, 2003. Its primary goal is to obtain data on and appropriately regulate some 30,000 chemicals that are produced or imported in excess of 1 metric ton per year and on which there is little toxicity and environmental information. In eliminating the practical distinction between "new" and "old" chemicals, it differs from the toxicity testing and risk assessment of chemicals under the U.S. Toxic Substances Control Act.

TABLE 6-1 REACH Testing Strategy

Test	1 metric ton or more	10 metric tons or more	100 metric tons or more	1,000 metric tons or more
Skin irritation or skin corrosion	R_a	R_a	R_a	R_a
In vivo skin irritation	NR	R_a	R_a	R_a
Eye irritation	R_a	R_a	R_a	R_a
In vivo eye irritation	NR	R_a	R_a	R_a
Skin sensitization	R_a	R_a	R_a	R_a
In vitro gene mutation in bacteria	R	R	R	R
In vitro cytogenicity in mammalian cells	R_b	R_a	R_a	R_a
In vitro gene mutation in mammalian cells	R_b	R_c	R_c	R_c
In vivo mutagenicity	P_a	P_a	P_a	P_a
Acute toxicity by oral route	NR	R_a	R_a	R_a
Acute toxicity by inhalation	NR	R_a	R_a	R_a
Acute toxicity by dermal route	NR	R_a	R_a	R_a
Repeated-dose toxicity	NR	NR	NR	P_b
Short-term repeated-dose toxicity (28 days)	NR	R_a	R_a	R_a

(Continued)

The REACH program has four major components (EU 2004). First, a registration process requires manufacturers and importers of chemicals to obtain relevant information on their substances and to use the data to manage chemicals more safely. Second, unnecessary testing is avoided by having government regulators evaluate industry testing proposals and check compliance with registration requirements; evaluation also enables agencies to investigate chemicals that potentially pose risks. Third, substances with properties of very high concern will be subject to authorization; applicants will have to demonstrate that risks are adequately controlled, and uses of high-risk substances may be authorized if socioeconomic benefits outweigh the risks and there are no suitable alternative substances or technologies. Fourth, reducing testing in vertebrate animals is encouraged, and toxicity testing is subject to data-sharing. The committee focuses its evaluation on the overall toxicity-

TABLE 6-1 *(Continued)*

Test	1 metric ton or more	10 metric tons or more	100 metric tons or more	1,000 metric tons or more
Subchronic toxicity (90 days)	NR	P_b	P_b	P_b
Long-term repeated-dose toxicity	NR	NR	NR	P_b
Reproductive toxicity	NR	R_a	P_b	P_b
Screening for reproductive and developmental toxicity	NR	R_a	R_a	R_a
Developmental toxicity	NR	R_a	R_a	R_a
Two-generation reproductive toxicity if 28-day or 90-day exposure indicates adverse effects on reproductive organs or tissues	NR	P_b	P_b	R_a
Two-generation reproductive toxicity	NR	NR	NR	P_b
Pharmacokinetics	NR	R	R	R
Carcinogenicity	NR	NR	NR	P_b

Abbreviations: NR, not required; R, required; R_a, required with possible exception; R_b, required if test 6 positive; R_c, required if test 6 and 7 negative; P_a, proposed if test 6 or 7 or 8 positive; P_b, proposed if appropriate.

testing strategy and is not commenting on the legal and regulatory procedural implementation of the legislation.

The legislation, if enacted, will require that all chemicals used in commerce at over 1 metric ton per year have basic toxicity and risk information generated within an 11-year period and that chemicals of very high concern be treated like drugs—only uses approved by government authorities would be permitted. The toxicity testing that would be required is outlined in Table 6-1. The exposure route is generally specified to be the one most relevant for potential human exposure.

Toxicity testing to be conducted is based primarily on the quantity of chemical manufactured or imported, with cutoffs at 1, 10, 100, and 1,000 metric tons. Toxicity-testing requirements for chemicals manufactured in quantities of 1-10 metric tons per year include only skin irritation or corrosion and sensitization, eye irritation, and several in vitro gene-mutation and cytogenicity studies. Positive results may trigger a hazard classification or additional testing. More substantial toxicity testing is required for substances manufactured or imported in quantities of 10 metric tons per year or more. Those chemicals must have a chemical safety assessment (CSA) as part of a registration dossier. In the CSA, the derivation of the hazard classification must be documented with the assessment of persistence, bioaccumulation, and toxicity (PBT) and the assessment of whether the substance is very persistent and very bioaccumulative (vPvB). The CSA also includes the derived no-effect levels and predicted no-effect concentrations.

The European Centre for the Validation of Alternative Methods (ECVAM) is involved in facilitating the implementation of the REACH program, which calls for the use of alternative testing strategies where feasible. ECVAM is directly targeting the animal tests to be replaced. For example, the REACH legislation calls for data on acute lethal toxicity of chemicals. ECVAM is validating a five-tier testing strategy that would reduce the number of animals being used for acute lethality. Stage 1 involves a quantitative structure-activity relationship (QSAR) model, stage 2 a basal cytotoxicity test (such as the 3T3 neutral red uptake test), stage 3 a computer model and a biotransformation test for metabolism, and stage 4 a cell-specific toxicity test and a prediction model. In each of the four stages, a finding of "very toxic" would indicate that no further testing is done. A fifth stage—in vivo testing—may be needed when the previous stage indicates the absence of significant acute lethal potential that should be confirmed.

Committee Comments on REACH Program

The REACH program is designed primarily to broaden coverage by ensuring that all chemicals that are manufactured in or imported into the EU will undergo at least some evaluation. In this strategy, tonnage cutoffs are used as rough surrogates for potential human exposure to the chemical agents, thereby triggering different levels of toxicologic scrutiny. Tonnage may be an initial rough surrogate, but additional information regarding potential for human exposure (for example, whether the chemical is an intermediate that has very low exposure) may also be helpful in guiding toxicity testing. Furthermore, classification of chemicals according to their PBT and vPvB may be a useful and simple adjunct to priority-setting and hazard assessment.

The committee made several general observations about the REACH program. First, REACH specifies categories of tests or end points that require evaluation at each level without being prescriptive about how the test will be done. Such an approach enhances flexibility but potentially affects comparability of results and indicates the need for the tests to be validated against each other. Second, as discussed above, the process is iterative with an evaluation step by government agencies and a data-sharing requirement to try to ensure that unnecessary or redundant animal testing is not performed. That step reflects an effort to decrease animal use in toxicity testing. Third, the REACH program focuses more on screening a large number of chemicals than on attaining the depth that is often needed for quantitative dose-response assessment. However, the REACH program does allow for greater depth of testing to be triggered on the basis of initial results. Lastly, the REACH program is more directly comparable with the voluntary U.S. high-production-volume screening program than with the proposals for improvements in the EPA RfD-RfC review or the ILSI-HESI proposals. REACH has the advantage of generating at least some toxicity data on chemicals that in the United States are not subject to testing requirements.

National Toxicology Program: Roadmap for the Future

Since its inception in 1979, the NTP has provided data for identifying the health hazards posed by environmental agents, and the data have been used extensively by regulatory agencies in risk assessment. The NTP's recently released Roadmap for the Future (NTP 2004) has emphasized several ways in which it expects to make substantial progress in the next few years. First, the NTP envisions that studies of gene and protein expression and of metabolic profiling in tissues of exposed animals will help to characterize responses to environmental agents and therefore lead to a better understanding of the mechanisms by which such agents alter disease incidence and progression. Second, the NTP is exploring the use of nonmammalian in vivo assays in toxicity testing. For example, one project, which is expected to be completed in 2007, involves the potential use of *Caenorhabditis elegans* as a screen for developmental, neurologic, and behavioral toxicity. Finally, increasing emphasis will be placed on developing alternative assays in which critical molecular and biologic events that occur early in the causation of adverse effects can be targeted, quantified, and placed within the toxicity-testing framework. The systems developed may help in setting priorities among substances for further study or in refining biologically based dose-response (BBDR) models.

In the Roadmap, NTP efforts are directed toward three main goals: refinement of traditional toxicology assays; development of rapid, mechanism-based predictive screens for environmentally induced diseases; and improvement in the overall utility of NTP toxicity-testing assays for public-health decisions. Those goals and the NTP's plan to train scientists conducting NTP studies in the practice of humane science indicate the NTP's commitment to reducing the pain and distress of animals and to decreasing the number of animals used in toxicity testing. The Roadmap describes the following current NTP research initiatives:

- Review existing protocols for carcinogenicity, reproductive toxicity, developmental toxicity, and immunotoxicity and make appropriate changes in the protocols to increase efficiency and to improve the quality of data generated. For example, the NTP has initiated a review of its cancer bioassay and is using public scientific workshops as one mechanism to receive input. The first workshop is focusing on stocks and strains used in the rodent cancer bioassay to seek scientific input on whether they should be continued or replaced or multiple strains should be used. Future workshops will address other cancer-bioassay design issues. Reviews of noncancer assays—such as developmental toxicity, immunotoxicity, and neurotoxicity studies—are expected within the next 3 years.

- Devise a strategy for routine incorporation of genomic analysis into current toxicity studies within the next year. The goal is to provide mechanistic information for improved understanding of toxicity. Resulting data on the identification of mechanistic targets would be used in the overall toxicity evaluation and in the development of high-throughput systems. The NTP expects such data also to permit the application of toxicity findings on a chemical to other chemicals that act by the same mechanisms. Data collection for some genomic analyses is already under way.

- Improve the use of pharmacokinetic information in toxicologic evaluations made by the NTP. The effort seeks to strengthen

NTP work in pharmacokinetics by establishing a minimal pharmacokinetic dataset on agents evaluated by the NTP, developing and validating PBPK models for agents of concern, building a multidisciplinary capacity within the NTP to do the work, and developing training and user-friendly tools.

- Continue to explore nonmammalian in vivo alternatives to toxicity testing as tools for medium-throughput screening (MTS) of specific toxicity end points.

- Expand the use of imaging technologies for detecting and quantifying molecular and cellular lesions and for improving the speed and precision of pathology reviews. This approach could increase the statistical power of NTP studies and reduce the number of animals used in the cancer bioassay and other studies. In the near term, the NTP plans to capture digital images from all pathology slides from the cancer bioassay routinely and to develop new image-analyzing techniques to guide the review and evaluation of lesions. A long-term goal is to provide a digital archive of such images from NTP studies that is accessible via the NTP's website.

Most of the above actions are intended to refine and increase the efficiency of the NTP's toxicity-testing strategy and study portfolio. The NTP Roadmap notes, however, the time-consuming and resource-intensive nature of traditional toxicity testing and the consequently large volume of newly introduced and existing chemicals in commerce that have been inadequately assessed for toxicity. The Roadmap outlines the NTP's long-range research strategy to develop rapid screening systems for providing toxicity information on large numbers of chemicals. The NTP plans to develop further MTS systems, such as in vivo nonmammalian systems, including *C. elegans*. High-throughput screening (HTS) may enable the evaluation of thousands of agents rapidly with in vitro biologic systems and robotics technology. The NTP's emphasis is on cellular targets known to influence disease or data interpretation. The NTP expects initially to target genotoxicity, cytotoxicity, cell proliferation, apoptosis, and some receptor-mediated activities and to run all relevant agents previously tested by the NTP through HTS with a set of other identified chemicals.

The databases created with the HTS and MTS can be used to determine the predictive value of the higher-throughput screens. In the short term, the NTP expects HTS to help to set priorities among agents for more extensive testing. In the longer term, the NTP hopes that such screens will become sufficiently developed and validated to be usable more directly in public-health decision-making. The NTP has staged its work in this field in phases, as outlined in Box 6-1.

Committee Evaluation of the NTP Roadmap

The NTP Roadmap is designed both to refine existing toxicity-testing approaches and to develop new approaches for eventual incorporation into toxicity testing. Although the NTP strategy has been presented publicly in bold brushstrokes rather than in detail, the general approaches are of great interest. The NTP's near-term efforts to refine and extend its toxicity tests and to improve the generation and use of pharmacokinetic and mechanistic data promise to increase the depth of toxicity information on chemicals assayed and to provide greater insight in applying test findings to humans. Furthermore, the NTP's stated objective of reducing animal use and increasing efficiency may be partly realized. However, as acknowledged by the NTP, the result-

---

BOX 6-1 Roadmap Activities: High-Throughput Screening (HTS)

Short-Term Activities
- Catalog available assays
- Convene working groups to provide advice on selection of assays
- Develop assays
- Identify initial set of chemicals for testing

Medium-Term Activities
- Continue assay development
- Validate individual assays
- Develop methods for analysis of data
- Develop HTS database
- Review effectiveness

Long-Term Activities
- Develop mechanisms to make chemical sets and tissue banks available for external researchers
- Evaluate HTS data for predictability of toxicity
- Develop a communication plan
- Review effectiveness

ing test portfolio would still be resource-intensive and incapable of addressing large numbers of chemicals that require some level of toxicity assessment. In the relative near term, the alternative—in vivo non-mammalian MTS—holds promise for improvements in cost, animal use, resources, and throughput, but it generally received little attention in the Roadmap. The extent of the NTP MTS effort is unclear. The NTP's development of such systems to the point where they may be used in public-health decision-making (for example, NTP identifications of chemicals as reproductive toxicants) would be a major advance. In the long term, HTS has the potential to increase the breadth of chemical assessment substantially while preserving the goal of limiting animal testing. Development of improved biomarkers of effect with new -omics approaches would contribute to all four goals of increasing breadth, increasing depth (if biomarkers provide valid mechanistic information), conserving animals, and reducing costs. A specific focus on developing nonanimal models by an agency, such as the NTP, is needed if such approaches are to become a useful alternative to traditional toxicity testing in animals.

COMMITTEE EVALUATION OF SUGGESTED IMPROVEMENTS IN TESTING STRATEGIES

Several government agencies and nongovernment organizations have recognized important gaps and inefficiencies in toxicity testing. EPA, ILSI-HESI, the NTP, and the EU have assessed or undertaken initiatives to build on and improve testing strategies. There is much to learn from their varied assessments and strategies.

EPA evaluated its testing requirements for pesticides and toxic substances in the context of establishing RfDs and RfCs, found substantial data gaps in life stages studied and end points assessed, and made recommendations for improvements in the near term by using existing toxicity-testing methods and techniques. ILSI-HESI convened a panel of scientists to evaluate EPA's approaches to toxicity testing for pesticides and developed its own set of recommendations. EPA made ambitious recommendations that would increase the breadth of toxicity testing in two key ways: coverage of different life stages and improved coverage of different end points, such as cardiovascular effects and immunotoxicity. In doing so, EPA explored alternative test protocols that would increase the complexity of some tests but ultimately conserve animal use while increasing critical toxicity information. ILSI-HESI emphasized potential test redundancies and test elimination. It recommended eliminating the mouse carcinogenicity study and 1-year dog study and making the rat acute toxicity study optional, thereby reducing animal use, the robustness of the database, and the potential to confirm findings. It also included proposals to increase the breadth of testing by modifying tests to evaluate more end points. It did not propose specific protocols to address effects of exposures in the elderly, although it did suggest that special studies could be triggered by findings in first-tier studies in young adult animals. Both EPA and ILSI-HESI recommended increased efforts to generate pharmacokinetic data, a practice already in place at the NTP, which increases the depth of available toxicity data. Both proposals also suggested reductions in group sizes in toxicity testing. Although in some cases the evaluation of end point or life stage may be more detailed, the small group size (for example, 20 vs 50 in the unified study and six vs eight dogs in the systemic-toxicity evaluations) can significantly reduce the statistical power of a study to detect effects.

The EU is engaged in a bold effort to restructure its approach to toxicity testing to screen the backlog of tens of thousands of industrial chemicals that have not been adequately assessed. It has taken the approach of attempting to broaden the screening of chemicals maximally while minimizing animal use in existing protocols. It is unclear whether the EU approach will provide adequate depth of information for dose-response assessment, especially with respect to the breadth of coverage of relevant end points and life stages. The approach to obtaining test data for detailed risk assessment is still under development, although for some chemicals (pesticides) schemes similar to those used in the United States are in place. It is therefore difficult to compare the REACH approach directly with the EPA and ILSI-HESI approaches because they are designed primarily to focus on different subsets of chemicals (tens of thousands of existing industrial chemicals vs hundreds of pesticides). The REACH approach includes greater depth of information and coverage of end points and life stages than the current EPA approach to testing of existing or new industrial chemicals under the Toxic Substances Control Act.

The NTP has not proposed specific changes in toxicity tests but has initiated a process, involving scientific workshops and public input, that could in the near term lead to fundamental changes in the design and conduct of the bioassays of cancer, reproductive and developmental toxicity, and other end points. The NTP is beginning by evaluating and considering changes in cancer-bioassay design. Also in the near term, the NTP seeks to improve testing by capitalizing on recent advances in imaging technologies, devising strategies for routine incorporation of toxicogenomic analyses, further developing nonmammalian test systems, and using transgenic animals. This effort goes considerably beyond the proposals and recommendations of ILSI-HESI and EPA. That is in part understandable in that some of the NTP effort involves technology transfer and the development and validation of new approaches; the changes are not yet sufficiently well defined to be incorporated into mandated test requirements, such as the EPA pesticide test guidelines. However, specific study designs and protocols should be given consideration for such purposes as they emerge from the NTP program. The NTP also is taking a longer view through its initiative to develop rapid screening systems for providing toxicity information on large numbers of chemicals.

The NTP approach is the only one that incorporates strategies that have potential for satisfying all four key goals of increasing breadth and depth and decreasing animal use and costs. Yet the NTP HTS of large numbers of chemicals relies in the long term on the development of new methods that are as yet unproved and will have to be evaluated. The NTP HTS approach is many years away from being practical for adoption in testing requirements by a regulatory agency, such as EPA. In the meantime, some of the near-term initiatives at the NTP, as well as strategies proposed in the EPA review and perhaps in the ILSI-HESI papers and in the REACH proposal, may be of use for improving current regulatory toxicity testing.

The committee identified several recurring themes and questions in the various reports that it was asked to review:

- *Which environmental agents should be tested?* All new and existing environmental agents should be evaluated; however, the intensity and depth of testing should be decided according to practical needs, including the use of the chemical, the likelihood of human exposure, and the scientific questions that such testing must answer to support a reasonable science-policy decision. Fundamentally, the design and scope of a toxicity-testing approach needs to reflect risk-management needs. Regulatory agencies are pursuing different testing strategies depending on whether their goal is generating detailed data for optimal dose-response assessment of pesticides or screening existing industrial chemicals. Hence, neither the EPA testing recommendations in its RfD-RfC review nor the EU REACH program is "correct" or "incorrect," because they are designed for different purposes.

•   *How should priorities for testing chemicals be set?* Priority-setting should be a key component of any testing strategy that is designed to address a large number of chemicals, and a well-designed scheme is essential for systematic testing of industrial chemicals on which there are few data. It makes sense to consider exposure potential in designing test strategies. Chemicals to which people are more likely to be exposed or to which some populations may receive relatively high exposures—whether they are pesticides or industrial chemicals—should undergo more in-depth testing. This concept is embedded in several existing and proposed strategies. In some strategies, production volume is the primary measure of potential human exposure. But production volume alone may not be the best surrogate of human exposure. Other important factors to consider are use, exposure patterns, and a chemical's environmental persistence and bioaccumulation, which is important because of the potential for increasing exposure over time and continuing exposure even after use has ceased. Indicators of potential toxicity from existing toxicity data or structural analogues and computational approaches, such as structure-activity relationships, may help to refine priorities further. In addition, there has been some investigation of high-throughput in vitro methods as a possible priority-setting tool. The committee will discuss possible priority-setting approaches in greater depth in its second report.

•   *What strategies for toxicity testing are the most useful and effective?* Existing test strategies include test batteries, tiered-testing strategies, tailored approaches, and strategies that combine various approaches. The committee finds that there are pros and cons of various approaches but leans toward tiered testing with the goal of focusing resources on the evaluation of the more sensitive adverse effects of exposures of greatest concern rather than full characterization of all adverse effects irrespective of relevance for risk-assessment needs. A tiered-testing approach would require that EPA have clear regulatory authority to require additional testing of pesticides or industrial chemicals beyond the first tier when it is indicated. Such a strategy would also require a priori stopping rules to prevent all chemicals from going through all tiers. For example, some observers have expressed the concern that some potentially "positive" result may often pop up in tiered testing and trigger the next tier of testing of nearly all chemicals, which would erase the benefit of tiering. Other observers have expressed the concern that in tiered strategies, strong positive results may simply trigger moving to ever-higher and more time-consuming tiers of testing while delaying risk-management action based on the findings. Both concerns are related to the need to have clear rules that move environmental agents out of the tiered-testing strategy into either a "stop" category or a preliminary risk-management category. One important point regarding the rules is that one should determine whether additional testing is likely to make a difference in human risk assessment. For example, the margin of exposure may be sufficiently large that further testing is not needed to make a decision or the end point in question is unlikely to be more sensitive than other end points of concern, in which case no further testing regarding that end point would be needed.

- *How can toxicity testing generate data that are more useful for human-relevant dose-response assessment?* Many observers have criticized existing approaches to toxicity testing on the grounds that it is difficult to use their results in risk assessment. In particular, observed results are often difficult to extrapolate with confidence to humans at environmentally relevant doses. As a result, such extrapolations are often made with little scientific justification, and conventional uncertainty factors are used to bridge the gaps.

Current approaches to toxicity testing could be enhanced in some cases by the use of pharmacokinetic data from basic ADME studies to derive dose information before embarking on toxicity testing and by the judicious use of some pharmacokinetic data to aid in extrapolation. In addition, adding lower doses to some protocols could help with obtaining more environmentally relevant information, although to maintain adequate statistical power increased group size at lower doses may be needed, with attendant increases in cost. Using benchmark dose extrapolations in place of NOAELs is another strategy that has been shown to be useful in maximizing the use of the existing dataset. However, the current menu of approaches to toxicity testing is unlikely to solve the fundamental problem. The committee cautions against indiscriminately generating large amounts of data on single chemicals in an effort to generate the optimal dose-response dataset. Such an approach entails the substantial use of animals and often comes at the expense of broader screening that may also be useful. Newer approaches to toxicity testing may help to address this problem and will be discussed in the committee's second report. In the interim, uncertainty factors must still be used to address scientific uncertainties.

• *How can toxicity testing be applied to a broader universe of chemicals, life stages, and end points?* It is clear that there are major gaps in current toxicity-testing approaches. The real importance of the gaps is a matter of debate and depends on whether effects of public-health importance are being missed by current approaches. Historically, an important end point or life-stage susceptibility has in many cases been missed in initial toxicity testing. However, it is neither practical nor desirable to attempt to test every chemical (or mixture) against every end point during a wide range of life stages. The committee recommends toxicity screening of every agent to which there is a strong potential for human exposure. A well-designed tiered strategy could help to set priorities among environmental agents for screening and could identify end points or mechanisms of action that would trigger more in-depth testing for various end points or in various life stages. Newer methods of screening that incorporate such tools as -omics or computational toxicology may help to screen more chemicals rapidly and trigger appropriate further testing where necessary. Those approaches are discussed briefly in Chapter 7 of this report and will be discussed in more depth in the committee's second report.

• *How can environmental agents be screened with the minimal use of animals and minimal expenditure of time and other resources?* One strategy that is useful to reduce animal use is the grouping of chemicals of similar structural class and the in-depth testing of only one or a few representative chemicals; risk assessments of all chemicals in the class would then be based on the resulting data. In grouping chemicals, known modes of action (for example, nicotinic agonists) should be emphasized. Such strategies should address any data needed to support application of study findings on the representative chemicals to other chemicals in the group. Newer approaches also have great promise. Chapter 7 discusses current developments in reduction, refinement, and replacement of animals in toxicity testing and newer technologies that hold promise for greatly reducing the reliance on animal testing.

• *How should tests and testing strategies be evaluated?* An important consideration in evaluating test strategies is the risk-management context in which they are being applied. For example, intensive study of untested industrial chemicals is of little use if it means that few chemicals can be addressed. However, further exploratory and intensive study of a relatively well-tested chemical can be of great value if costs of controlling exposure are high, there is widespread and relatively high exposure, and study is likely to refine the risk assessment substantially. Test strategies may be evaluated in terms of the value of information they provide in light of the cost of the testing and the animal resources expended. That should be kept in mind in considering testing strategies in terms of the four testing objectives—increasing depth of knowledge for more accurate risk assessment; increasing coverage of chemicals, life stages, and end points; preserving animal welfare; and minimizing cost.

In evaluating new tests and test strategies, there remains the difficult question of what is to serve as a "gold standard" for performance. Simply comparing the outcomes of new tests with the outcomes of current tests may not be the best approach; whether it is will depend on the reliability and relevance of the current tests. Another consideration is how test results will be used in the assessment. Even if a test strategy provides robust and informative data, the risk assessor may be unable, because of legal constraints or risk-assessment guidelines, to use the data. Ideally, regulations and risk-assessment guidelines will evolve with testing capabilities and scientific understanding. That issue will increase in importance with greater use of screening approaches that produce indirect evidence (in vitro tests, gene arrays, and mode-of-action screens), for both cancer and noncancer end points.

REFERENCES

Cooper, V.L., J.C. Lamb, S.M. Barlow, K. Bentley, A.M. Brady, N.G. Doerrer, D.L. Eisenbrandt, P.A. Fenner-Crisp, R.N. Hines, L.F.H. Irvine, C.A. Kimmel, H. Koeter, A.A. Li, S.L. Makris, L.P. Sheets, G.J.A. Speijers, and K.E. Whitby. 2006. A tiered approach to life stages testing for agricultural chemical safety assessment. Crit. Rev. Toxicol. 36 (1):69-98.

Doe, J.E., A.R., Boobis, A. Blacker, V.L. Dellarco, N.G. Doerrer, C. Franklin, J.I. Goodman, J.M. Kronenberg, R. Lewis, E.E. McConnell, T. Mercier, A. Moretto, C. Nolan, S. Padilla, W. Phang, R. Solecki, L. Tilbury, B. van Ravenswaay, and D.C. Wolf. 2006. A tiered approach to systemic toxicity testing for agricultural chemical safety assessment. Crit. Rev. Toxicol. 36 (1):37-68.

EPA (U.S. Environmental Protection Agency). 1998a. Health Effects Test Guidelines OPPTS 870.1100. Acute Oral Toxicity. EPA 712-C-98-190. Office of Prevention, Pesticides, and Toxic Substances, U.S. Environmental Protection Agency, Washington, DC [online]. Available: http://www.epa.gov/opptsfrs/OPPTS_Harmonized/870_Health_Effects_ Test_Guidelines/Series/870-1100.pdf [accessed March 15, 2005].

EPA (U.S. Environmental Protection Agency). 1998b. Health Effects Test Guidelines OPPTS 870.1200. Acute Dermal Toxicity. EPA 712-C-98-192. Office of Prevention, Pesticides, and Toxic Substances, U.S. Environmental Protection Agency, Washington, DC [online]. Available: http://www.epa.gov/opptsfrs/OPPTS_Harmonized/870_Health_Effects_ Test_Guidelines/Series/870-1200.pdf [accessed Oct. 7, 2005].

EPA (U.S. Environmental Protection Agency). 1998c. Health Effects Test Guidelines OPPTS 870.1300. Acute Inhalation Toxicity. EPA-712-C-98-193. Office of Prevention, Pesticides, and Toxic Substances, U.S. Environmental Protection Agency, Washington, DC [online]. Available: http://www.epa.gov/opptsfrs/OPPTS_Harmonized/870_Health_Effects_ Test_Guidelines/Series/870-1300.pdf [accessed Oct. 7, 2005].

EPA (U.S. Environmental Protection Agency). 1998d. Health Effects Test Guidelines OPPTS 870.4300. Combined Chronic Toxicity/ Carcinogenicity. EPA 712-C-98-212. Office of Prevention, Pesticides, and Toxic Substances, U.S. Environmental Protection Agency, Washington, DC [online]. Available: http://www.epa.gov/opptsfrs/OPPTS _Harmonized/870_Health_Effects_Test_Guidelines/Series/870-4300.pdf [accessed March 15, 2005].

EPA (U.S. Environmental Protection Agency). 1998e. Health Effects Test Guidelines OPPTS 870.3800. Reproduction and Fertility Effects. EPA 712-C-98-208. Office of Prevention, Pesticides, and Toxic Substances, U.S. Environmental Protection Agency, Washington, DC [online]. Available: http://www.epa.gov/opptsfrs/OPPTS_Harmonized/870_ Health_Effects_Test_Guidelines/Series/870-3800.pdf [accessed March 15, 2005].

EPA (U.S. Environmental Protection Agency). 1998f. Health Effects Test Guidelines OPPTS 870.3100. 90-Day Oral Toxicity Study in Rodents. EPA 712-C-98-199. Office of Prevention, Pesticides, and Toxic Substances, U.S. Environmental Protection Agency, Washington, DC [online]. Available: http://www.epa.gov/docs/OPPTS_Harmonized/870 _Health_Effects_Test_Guidelines/Series/870-3100.pdf [accessed Oct. 26, 2005].

EPA (U.S. Environmental Protection Agency). 1998g. Health Effects Test Guidelines OPPTS 870.2400. Acute Eye Irritation. EPA 712-C-98-195. Office of Prevention, Pesticides, and Toxic Substances, U.S. Environ-mental Protection Agency, Washington, DC [online]. Available: http://www.epa.gov/opptsfrs/OPPTS_Harmonized/870_Health_Effects_ Test_Guidelines/Series/870-2400.pdf [accessed April 7, 2005].

EPA (U.S. Environmental Protection Agency). 2002. A Review of the Reference Dose and Reference Concentration Processes. Final Report. EPA/ 630/P-02/002F. Risk Assessment Forum, U.S. Environmental Protection Agency, Washington, DC [online]. Available: http://www.epa. gov/iris/RFD_FINAL%5B1%5D.pdf [accessed March 11, 2005].

EPA (U.S. Environmental Protection Agency). 2005. OPPTS Harmonized Test Guidelines. Series 870 Health Effects Test Guidelines-Final Guidelines. Office of Prevention, Pesticides, and Toxic Substances, U.S. Environmental Protection Agency, Washington, DC [online]. Available: http://www.epa.gov/opptsfrs/OPPTS_Harmonized/870_Health_Effects_ Test_Guidelines/Series/ [accessed May 5, 2005].

EU (European Union). 2004. The REACH Proposal Process Description. REACH (Registration, Evaluation and Authorisation of Chemicals), Enterprise and Industry, European Union [online]. Available: http:// europa.eu.int/comm/enterprise/reach/docs/reach/reach_process_ description-2004_06_15.pdf [accessed March 8, 2005].

ILSI-HESI (International Life Sciences Institute Health and Environmental Sciences Institute). 2004a. Systemic Toxicity White Paper. Systemic Toxicity Task Force, Technical Committee on Agricultural Chemical Safety Assessment, ILSI Health Sciences Institute, Washington, DC. November 2, 2004.

ILSI-HESI (International Life Sciences Institute Health and Environmental Sciences Institute). 2004b. Life Stages White Paper. Life Stages Task Force, Technical Committee on Agricultural Chemical Safety Assessment, ILSI Health Sciences Institute, Washington, DC. November 2, 2004.

ILSI-HESI (International Life Sciences Institute Health and Environmental Sciences Institute). 2004c. The Acquisition and Application of Absorption, Distribution, Metabolism, and Excretion (ADME) Data in Agricultural Chemical Safety Assessments. ADME Task Force, Technical Committee on Agricultural Chemical Safety Assessment, ILSI Health Sciences Institute, Washington, DC. November 2, 2004.

NTP (National Toxicology Program). 1984. Report of the NTP Ad Hoc Panel on Chemical Carcinogenesis and Testing Evaluation. NTP Board of Scientific Counselors, U.S. Department of Health and Human Services. August 17, 1984 [online]. Available: http://ntp-server.niehs.nih. gov/files/NTPAdHocPanel84.pdf [accessed Oct. 26, 2005].

NTP (National Toxicology Program). 2004. The NTP Vision for the 21st Century. National Toxicology Program, National Institute for Environmental Health, Research Triangle Park, NC [online]. Available: http://ntp-server.niehs.nih.gov/ntp/main_pages/NTPVision.pdf [accessed March 11, 2005].

NTP (National Toxicology Program). 2005. Presentations at the Workshop: Animal Models for the NTP Rodent Cancer Bioassay: Strains and Stock- Should We Switch? June 16-17, 2005, National Institute of Environmental Health Sciences, Research Triangle Park, NC [online]. Available: http://ntp-server.niehs.nih.gov/files/Agenda_Presentations.pdf [accessed Oct. 26, 2005].

7

Alternative Approaches and Emerging Technologies

There is a coordinated international effort to develop alternatives to animals for toxicity testing of environmental agents. Numerous methods have already been developed and validated to reduce, replace, or refine animal testing, and many more are under development in the United States and Europe. The effort to reduce animal use has generated some additional benefits. For example, some nonanimal methods provide useful mechanistic information that can offer insight into the likely human relevance of observed findings or may offer the ability to predict patterns of toxicity. Furthermore, some approaches that use alternative non-mammalian species allow testing of much larger numbers of organisms, thereby increasing statistical power for evaluating dose-response relationships at the low end of the curve.

This chapter reviews approaches specifically focused on alternatives to animal testing that reduce, replace, or refine animal use. The second part discusses some new toxicity-testing approaches (-omics technologies and computational toxicology) that may have longer-term potential for achieving greater depth, breadth, animal welfare, and conservation in toxicity testing. The chapter concludes with a discussion of validation to emphasize the importance of evaluating new toxicity-testing methods to ensure that the information obtained from them is at least as good as, if not better than, conventional mammalian models. Validation, as defined in this chapter, is a formal process that grew out of the experience of the European Centre for the Validation of Alternative Methods (ECVAM), the Interagency Coordinating Committee on the Validation of Alternative Methods (ICCVAM), and others in evaluating the performance of new tests. The important point for this report is that validation is now seen as a formal, although flexible, process that new tests must satisfy to be accepted by regulators. The details of validation exercises may vary as one shifts from in vitro and in vivo tests to -omics and computational toxicology techniques.

ALTERNATIVES TO CURRENT
ANIMAL-TESTING APPROACHES

One of the tensions in designing new chemical-testing strategies is between reducing animal use and suffering and regulatory needs for more information on a wider array of chemicals or more detailed information on a smaller group of chemicals. Russell and Burch (1992) provided a framework for addressing that tension. They proposed that scientists pursue techniques and approaches that follow the Three Rs, namely, methods that can *r*eplace or *r*educe animal use in specific procedures or *r*efine animal use to eliminate or decrease animal suffering. Replacement, reduction, and refinement have also come to be known as alternative methods.

First proposed in 1959, the Three Rs approach (3Rs) advanced in the 1980s when cosmetics and consumer-product companies began to invest millions of dollars in alternative methods in response to consumer pressure (Stephens et al. 2001). During that same decade, national governments incorporated the Three Rs approach into their animal-protection legislation and in some cases began to fund research on and development of alternatives, academic centers devoted to the alternatives began to be established, the field of in vitro toxicology blossomed, and companies began to market alternative test kits. In the 1990s, government centers devoted to the validation and regulatory acceptance of alternative methods were established in Europe and the United States, alternative tests began to be formally approved and accepted by regulatory agencies, and the triennial World Congresses on Alternatives were inaugurated. There is evidence that, owing in part to the implementation of Three Rs approaches, use of laboratory animals in research and testing in the United States decreased by about 30%[1] in the decade after the estab-

---

[1]Estimate based on comparison of average number of Animal Welfare Act (AWA)-covered animals used per year in 1994-2003 and average number of AWA-covered animals used per year in 1984-1993. Source: Animal Welfare Reports, USDA/APHIS.

lishment of ICCVAM in 1997, which marked the beginning of widespread efforts to implement the Three Rs. In the 21st century, as acceptance and implementation of the Three Rs approach continue to spread, a major challenge in advancing the approach is to harness the potential of new technologies, including -omics, to replace, reduce, and refine animal use.

The following sections explore in more detail the refinement, reduction, and replacement alternatives. The replacement of commonly used laboratory animals with less sentient animal species is addressed specifically.

Refinement Alternatives

Refinement alternatives are changes in existing practices that either decrease animal pain and distress or increase animal welfare. Refinements are best practices, namely, ways of carrying out animal-based procedures and practices that ensure the best practical outcomes with respect to both animal welfare and science. The principle of refinement can be applied to any aspect of laboratory care and use—including anesthesia, analgesia, supportive veterinary care, and euthanasia—and to the more general aspects of animal transport, handling, housing, environmental enrichment, and personnel training (Morton 1995). Refinement approaches of particular relevance to toxicity testing include best practices in dose administration, dose-volume limits, and humane end points (Hendriksen and Morton 1999; ILAR 2000; OECD 2000; Diehl et al. 2001; Stephens et al. 2002). Humane end points in an animal experiment are early indicators of pain, distress, or death and, once validated, can be used to terminate an experiment early to preclude or lessen animal suffering without compromising study objectives (Stokes 2000). The application of humane end points is often associated with frequent monitoring of animals and scoring of their clinical signs. Scoring systems are an important tool for evaluating the efficacy of proposed refinements.

In toxicology, refinements include not only modifications of existing tests but also new animal-based tests that result in less pain or distress than conventional procedures or in no pain or distress. For example, historically the guinea pig maximization test (GPMT) was the conventional assay for acute contact dermatitis (ACD). A new procedure, the local lymph node assay (LLNA), assesses ACD by examining local lymph node proliferation instead of the ensuing clinically evident allergic reaction. The animals, in this case mice, are euthanized before experiencing the discomfort of ACD. The LLNA can be considered an elaboration of the humane-end-point approach, which was made possible by knowledge of the mechanism of ACD. The LLNA has been accepted by EPA, the Food and Drug Administration (FDA), and the Occupational Safety and Health Administration (OSHA) as a refinement alternative to the GPMT for assessing ACD (NTP 1999).

Refinements in toxicology obviously benefit the animals involved in testing, but they can also be advantageous from scientific and societal viewpoints. Pain or distress stemming from poor technique can cloud study outcomes (Morton et al. 2001). Refined approaches, such as the use of humane end points, can lead to earlier completion of testing. Scoring of clinical signs can reveal toxicologic outcomes that might have been overlooked if death were the only outcome noted. Finally, implementing refinement can improve the morale of laboratory personnel and help to satisfy mandates in humane legislation, such as the U.S. Animal Welfare Act, with its emphasis on minimizing pain and distress.

Reduction Alternatives

Reduction alternatives are methods that use fewer animals than conventional procedures but yield comparable levels of information. They can include methods that use the same number of animals but yield more information so that fewer animals are needed to complete a given project or test (Balls et al. 1995). One of the most dramatic illustrations of reduction is the acute systemic toxicity-testing guidelines of the Organisation for Economic Co-operation and Development (OECD), which apply primarily to industrial chemicals. The number of animals used in OECD's Test Guideline 401 for the LD_{50} test dropped from 100 to 25 when the guideline—adopted in 1981—was modified in 1987. OECD also adopted three new guidelines in the 1990s that reflected additional reduction approaches that typically use under 10 animals per test. The new alternatives—the up-and-down procedure, the fixed-dose procedure, and the acute-toxic-class method—led OECD to drop Guideline 401 altogether from its guidelines in 2002 (OECD 2002a).

One straightforward way to explore reduction approaches for a given animal test is through retrospective analyses of test data on individual animals. If N is the number of animals conventionally used in a test, do (N − 1) animals typically yield the same conclusion? Do (N − 2) animals yield the same conclusion? That approach has been applied to the Draize eye-irritancy test to reduce the conventional number of animals used per test from six to three (see EPA 1998).

A rigorous application of experimental design and statistical approaches is one of the best ways to pursue reduction in animal numbers (Festing et al. 1998; Vaughan 2004). Statistical aids can yield precise estimates of the number of animals needed to test a hypothesis. Block designs can lead to reduction in animal numbers. And using animals that have genetically defined backgrounds can limit statistical variance and thereby achieve a given level of statistical power with fewer animals (Russell and Burch 1992; Festing 1999).

Animal reduction can also be achieved by applying adaptive Bayesian statistical techniques to study design. Such approaches have been used in clinical trials for evaluating new drugs and have resulted in reduced numbers of subjects and early termination in specific arms of clinical trials, reducing ineffective treatments and life-threatening side effects and improving survival (Berry et al. 2002; Giles et al. 2003). The same techniques could be adapted to reduce the numbers of animals used in toxicity testing.

Various noninvasive imaging techniques can be used to track the progression of toxic effects or disease in a cohort of animals, eliminating the need for interim killing of animals at selected times. To date, those techniques, such as biophotonic imaging (Contag et al. 1996), have been implemented primarily in biomedical research, as opposed to toxicity testing. If applied to regulatory toxicity testing, they could not only reduce animal numbers in some tests but facilitate the refinement of tests by allowing the monitoring of animals over time to gauge how close they are getting to specified humane end points, such as tumor size.

Animal use can also be limited by careful design of testing schemes. For example, EPA modified the testing scheme in its high-production-volume (HPV) chemical testing program after pressure from animal protectionists. The agency called on program participants to take a number of steps intended to reduce animal use, including grouping chemicals into appropriate categories and testing only representative chemicals from a category, avoiding some types of testing of closed-system intermediates, and encouraging a thoughtful, qualitative analysis rather than a rote checklist approach (see EPA 1999). Reducing animal numbers in toxicity tests not only subjects fewer animals to potential suffering but has the potential to lower the cost of testing.

Replacement Alternatives

Replacement alternatives use nonanimal approaches in lieu of animal-based methods. In toxicology, such nonanimal approaches include physiochemical measures, quantitative structure-activity relationship (QSAR) models, and other methods. Replacement might include substituting invertebrates in testing typically done with vertebrates, for example, the use of *Caenorhabditis elegans* in chronic toxicity testing. It might also include substituting primary culture of tissues or cells, such as neuromuscular preparations, for whole animals; however, such cultures entail animal use to harvest the tissues that will be cultured and therefore do not truly replace animal use.

Some nonanimal methods can serve as screens to limit the number of chemicals that move on to later stages of testing. For example, a simple pH determination can characterize a chemical as highly acidic or alkaline and so almost certainly an eye irritant, thus obviating a Draize eye-irritancy test in rabbits (see OECD 2002b). Such a screen can be labeled a "partial replacement" to distinguish it from a nonanimal method that serves as the definitive test, a "full replacement." Full replacements clearly are more satisfactory from a humane perspective, but partial replacements do limit animal use and suffering in toxicity tests.

Replacement approaches have been successfully implemented over the last several decades for a variety of applications, including culturing viruses, assaying vitamins, diagnosing pregnancy, and preparing monoclonal antibodies (Stephens 1989). In toxicology, in vitro tests have shown great potential as replacement alternatives. The Ames mutagenesis test, developed in 1971, was the first in vitro test used in regulatory toxicology. In vitro tests and other nonanimal methods have since been accepted in regulatory toxicology case by case after the development of the field of validation and the establishment of ICCVAM and ECVAM in the 1990s. In recent years, ICCVAM, ECVAM, and OECD have validated or accepted as validated a number of in vitro tests (see Chapter 2, Table 2-3), including the 3T3 neutral red update phototoxicity test, a skin-absorption assay, cytotoxicity assays for acute systemic toxicity, and skin-corrosivity assays, such as the transcutaneous electrical resistance assay, the Corrositex assay, and the Episkin and Epiderm assays (ICCVAM 2004; ECVAM 2005). Their validations have established the strengths and weaknesses of the assays and in some cases limited their applicability to particular chemical classes or levels within tiered testing strategies.

Overall, nonanimal approaches offer several potential advantages over classical animal-based tests. First, they can be less time-consuming and more humane. Second, they can be more mechanistically relevant to human toxicity when they are selected or tailored to reflect a specific biochemical pathway or a chemical receptor that does not occur in a given animal model. Third, they can allow for higher throughput. Because of the technical advantages, such approaches are being evaluated for large-scale testing programs, including HPV chemical testing and endocrine-disruptor testing. As the large-scale testing programs are developed and implemented, nonanimal methods are being incorporated as screens into tier-testing approaches with animal testing being reserved primarily for the highest tiers. Efforts to develop a full array of nonanimal methods to address all end points in some testing programs are under way (Worth and Balls 2002). That approach would rely on mechanistically based assays and, where appropriate, incorporate metabolic activation. Such an approach to toxicity-testing programs might be able to eliminate the need for extrapolation from animals to humans in some cases and to aid in identifying hazards to potentially sensitive human populations.

Use of Alternative Species

Nonmammalian vertebrates, such as fish, are being used increasingly in human health effects testing. To the extent that such species are less sentient than mammals, their use constitutes an example of refinement. Some nonmammalian species have a high degree of structural and physiologic similarity to higher vertebrates, enhancing the likelihood that similar toxicities would be produced. In addition, nonmammalian species have shorter developmental periods and shorter overall life spans, which are useful characteristics for simulating effects of chronic exposure. And they usually require simpler, less expensive laboratory maintenance than mammals.

The effectiveness of alternative models is well illustrated by historically prominent studies that used rainbow trout as a model for carcinogenicity and mechanistic cancer research. Trout have been shown to share many mechanisms of carcinogenesis with mammals, such as pathways of metabolic activation and production of mutagenic DNA adducts. Recently, the low cost and ease of maintenance of trout were taken advantage of to carry out the largest dose-response study of chemical-induced carcinogenesis ever conducted (William et al. 2003). The goal of this project, which used 42,000 trout, was to identify the dose at which one additional cancer in 10^3 animals occurred, an order-of-magnitude increase in sensitivity over the largest mouse study, which used 24,000 mice (Gaylor 1980). The dose-response data deviated significantly from linearity, although a threshold dose could not be statistically established. Studies that use large numbers of animals and thereby have increased sensitivity would have profound implications for modeling human health risk assessment if the animal models used were found to be relevant to humans.

Another fish model that is gaining increased attention from toxicology researchers is the zebrafish (Sumanas and Lin 2004). Zebrafish have many features that make them highly desirable as a laboratory model, including small size, high fecundity, and rapid development. The embryos are transparent, and this allows visualization of fundamental developmental processes with a simple dissecting microscope. A generation time of only 3 months makes genetic screening practical. Furthermore, a variety of diseases have been successfully modeled in zebrafish via simple genetic alterations or mutations. Much of the zebrafish genome has been sequenced, and at least two zebrafish oligonucleotide microarrays are available, each containing over 14,000 unique sequences. Transgenic zebrafish that express green fluorescent protein (GFP) under the control of various tissue-specific promoters have also been developed.

From a toxicology perspective, zebrafish have been shown to express the aryl hydrocarbon receptor (AhR) and the AhR nuclear translocator (ARNT), two proteins that are responsible for initiating the toxic effects of 2,3,7,8-tetrachlorodibenzo-*p*-dioxin (TCDD) and structurally related halogenated aromatic hydrocarbons in mammals (Andreason et al. 2002). Zebrafish respond to TCDD with induction of cytochrome P4501A, a key gene controlled by TCDD-activated AhR in all species examined (Andreason et al. 2002). Scientists at the National Toxicology Program (NTP) are evaluating zebrafish to determine their usefulness in screening chemicals for potential toxicity and carcinogenicity. Because of their genetic uniformity and low rates of spontaneous tumor, the use of zebrafish minimizes the experimental variability normally associated with other alternative animal species.

Although nonmammalian models show great promise at both ends of the toxicity-testing spectrum (screening and mechanistic studies), there are obvious limitations on the use and applicability of such nonmammalian species in some aspects of toxicity testing. Metabolic differences may be greater between nonmammalian species and humans than between humans and other mammals, so the use of such data for human health risk assessment may be more tenuous. Substantial anatomic and physiologic differences between mammals and other species will also prevent their application to assessment of some toxic end points.

EMERGING TECHNOLOGIES

Novel -omics technologies and computational toxicology may one day contribute to resolution of much of the current tension around the objectives of toxicity testing. The new fields are developing rapidly, and their integration into traditional testing strategies is being investigated. This section provides an overview of the tools, techniques, and science that show promise for advancing toxicity testing and risk assessment.

Genetics

Individuals differ in their responses to environmental toxicants, and that variability can be attributed to many factors. One possible factor is the variation in the human genome. Each person's genome is different, and the differences are thought to influence a person's response and susceptibility to a chemical exposure. The Human Genome Project at the National Institutes of Health has greatly facilitated the search for susceptibility genes—genes that influence a person's response to a stimulus or probability of developing a particular disease. In the last decade, researchers have been successful in identifying genes for diseases, such as cystic fibrosis, that are due to mutations in single genes. The effect of such a mutation is large and therefore relatively easy to identify. Identifying the susceptibility genes for complex human traits has been more challenging, but recent molecular and statistical advances stimulated by the Human Genome Project have led to the identification of susceptibility genes for several complex human diseases, such as asthma and Crohn's disease. Those advances have also led to identification of genetic variations that make some people more and other people less susceptible to environmental toxicants.

Several key developments in addition to the Human Genome Project have advanced the field of genetics. First is the characterization of DNA-sequence polymorphisms, particularly the single-nucleotide polymorphisms (SNPs).[2] Three entities—the SNP Consortium (TSC), the International HapMap Project, and the National Institute of Environmental Health Sciences (NIEHS) Environmental Genome Project (EGP)—have identified and characterized millions of SNPs. Specifically, they have provided positional information and allele frequencies and have developed assays for genotyping them. The SNPs identified by TSC and the HapMap Project are distributed across the entire genome and were not selected specifically for their functional significance. The SNPs identified by the EGP reside in environmentally responsive genes, such as genes involved with the cell cycle, DNA repair, and metabolism. The work of all three entities has provided a well-characterized set of SNPs that can be used as genetic "landmarks" to localize genes that influence one's susceptibility to disease and sensitivity to toxicants.

The second advance is the development of technologies for high-throughput genotyping. Although millions of polymorphisms have been identified, genotyping them for routine analysis has been an expensive, labor-intensive task. Until recently, genotyping was performed marker by marker; thus, the throughput was low and the cost high. Several recent developments allow thousands of markers to be genotyped in parallel. Large numbers of genotypes can be generated from DNA samples from many individuals. That advance is particularly important because the effect of each sequence variation is likely to be small, and these small effects would be very difficult to detect without a sample of adequate size.

The third advance is the improvement of phenotyping methods. A phenotype is the biochemical, physiological, or physical characteristics of an individual as determined by his or her genetic background and the environment. Defining phenotypes and collecting material for study often present challenges. To determine the genetic basis of a phenotype, one must study how the phenotype is passed along in families; therefore, phenotypic measurements and DNA from family members are often needed for analysis. However, in trying to define a phenotype that would indicate susceptibility to an environmental toxicant, it is difficult or impossible, to identify family members who have been exposed to the same

---

[2]Variations in DNA sequence that occur in more than 1% of the population are considered polymorphisms, and SNPs are polymorphisms that differ by one nucleotide.

agents under similar circumstances. To circumvent that problem, cell cultures from family members are exposed to environmental agents. The agent and dose are controlled, and a large amount of family material can be evaluated. Those studies have demonstrated that gene expression and phenotypes, such as cellular functions, can be accurately measured in cultured cells (Schork et al. 2002; Yan et al. 2002; Lo et al. 2003). In addition, the phenotypes identified in cultured cells are amenable to genetic analysis (Schadt et al. 2003; Greenwood et al. 2004; Morley et al. 2004). In recent studies, cells from members of large three-generation families were exposed to chemotherapeutic agents, such as cisplatin (Dolan et al. 2004), 5-fluorouracil, and docetaxel (Watters et al. 2004), and the genes that influence chemotherapy toxicity were identified (Dolan et al. 2004; Watters et al. 2004). Improvements in phenotyping methods are important for elucidating the genetics of chemical response and measuring the consequences of genetic variation.

Genomics

The human genome has been estimated to consist of about 25,000 genes. The gene-expression pattern varies from cell to cell and determines the identity of each cell. Cells induce or repress particular genes in response to environmental stimuli. Changes in gene expression help the cells to adapt to the "new" environment or repair damage resulting from the stimuli. One can identify genes that change in response to exposure by comparing the expression level of genes at baseline to the expression level in response to stimuli. With such technologies as microarrays, the expression levels of tens of thousands of genes can be measured accurately and efficiently. Those genes may serve as biomarkers of exposure and also aid in understanding the mechanism of action of the stimuli and the cellular pathways involved in the response.

Several groups, such as ILSI-HESI, have initiated projects to investigate the use of genomic data in risk assessment (Pennie et al. 2004; Hood 2004). Other organizations have initiated programs to investigate the use of genomic and other -omic technologies in toxicology. For example, the NRC Standing Committee on Emerging Issues and Data on Environmental Contaminants, which was convened at the request of NIEHS, currently is focused on toxicogenomics and its applications in environmental and pharmaceutical safety assessment, risk communication, and public policy.[3] The standing committee is not tasked with developing a consensus report but has convened workshops on a variety of topics, including use of bioinformatics to manage toxicogenomics information across laboratories; identification of critical knowledge gaps in cancer risk assessment and potential application of toxicogenomics technologies to address those gaps; identification of methods for communicating toxicogenomics information with the public and other nonexpert audiences; application of toxicogenomics to cross-species extrapolation to determine whether the effects of chemicals in animals can be used to predict human responses; and investigation of strategies to overcome obstacles to sharing toxicogenomics data.

Genomic experiments have been performed to answer questions in toxicology and include the following studies:

• Studies of transcriptional changes in cells exposed to a particular agent, such as the exposure of mouse hepatoma cells to chromium (Wei et al. 2004), the exposure of the liver of transgenic mice that have constitutively active dioxin-AhR to *N*-nitrosodiethylamine (Moennikes et al. 2004), exposure of MCF-7 cells to estrogen (Terasaka et al. 2004), and exposure of human keratinocytes to inorganic arsenic (Rea et al. 2003).

• Studies of dose-response assessment, such as those to evaluate dose-dependent expression changes in kidney HEK293 cells exposed to arsenite (Zheng et al. 2003).

• Studies of the extent of individual variation in response to exposure, such as those to determine susceptibility genes for resistance to dichlorodiphenyltrichloroethane (DDT) in *Drosophila* (Daborn et al. 2002; Pedra et al. 2004).

Most of the published studies focus on identifying the transcriptional changes associated with exposure. However, studies need to examine mechanisms of action of toxicants and determine general and toxicant-specific cellular responses. Furthermore, studies need to include sufficiently large samples, assess dose-response relationships, characterize the temporal nature of gene expression in relation to the relevant end point, and, to the extent possible, examine an appropriate variety of tis-

---

[3]Toxicogenomics uses genomic and other -omic technologies to study the genetic response to environmental pollutants or toxicants and ultimately to identify environmental agents that could cause adverse effects.

sues from different organisms. To identify susceptible populations and the potential hazards of chemicals with genomic technologies, large amounts of information must be accumulated and compared. Common microarray techniques and databases are being developed to make data storage and analysis feasible on a large number of experiments. Minimum Information About a Microarray Experiment for Toxicogenomics (MIAME/TOX) (EBI 2005) is a standard for microarray experiments; it is based on a particular microarray model and data-exchange format, and it allows integration of data from other sources, such as clinical data and data from histopathologic studies. That standard is intended to address at least some of the difficulties that arise in comparison of datasets that have been acquired with different technologies, compiled at different times, or generated from different laboratories and should facilitate the construction of databases with broader utility, such as those being developed by the National Center for Toxicogenomics.

Although genomics technologies, such as transcript profiling, have considerable potential in both predictive and mechanistic toxicology, their appropriate application to a risk-benefit analysis of novel chemical entities requires an understanding of the utility of the resulting data and, ideally, regulatory guidance or policy regarding their use. Recognizing the potential of genomics approaches, a number of regulatory agencies, including FDA and EPA, have issued draft guidance on the integration of these approaches into established risk-assessment schemes. For example, in 2005, FDA released final guidance on pharmacogenomics-data submission (FDA 2005) that recognizes the research applications of genomics, such as priority-setting among chemicals in a chemical class and selection of compounds for further development. Submission of genomic data is not required except when "known" or "probable" valid biomarkers of effect are recognized. In the absence of those biomarkers, data are required only for submission in an investigational new drug or new drug application filing if they are being used to support a safety argument (for example, the relevance of an effect in humans vs animal species), as a component in clinical trial design (for example, as a method to stratify patients or to monitor patients during the trial), or to clarify a labeling issue. FDA is also seeking voluntary submission of genomic data to increase its experience in handling and interpreting the data.

In a similar vein, EPA (2002) issued a brief interim policy statement on genomics technology that recognizes that genomics data would most likely be used as supportive or research data—that is, potentially used in ranking chemicals for further testing or in supporting regulatory action. In March 2004, EPA reviewed the potential role of genomics technologies across a broad array of issues related to toxicity testing, risk assessment, and regulation of chemicals in the environment (EPA 2004). The review was the product of the EPA Genomics Task Force, formed at the request of the EPA Science Policy Council.

The task force recognized that standardization of experimental design and the emergence of data-quality standards may be necessary for the use of the data in regulatory policy and processes. It identified four elements of regulation in which genomics activities are likely to influence regulatory practice, policy, or review: priority-setting among contaminants and contaminated sites, monitoring, reporting provisions, and enhanced risk assessment. Many research needs and activities were recognized—for example, linking the Office of Research and Development's Computational Toxicology Research Program to genomics activities, developing an analytical framework and acceptance criteria for genomics data, and developing internal expertise and methods to evaluate such data at EPA. Throughout the review, EPA recognized the role of the emerging technologies in informing the risk-assessment process and in potentially increasing the scientific rigor of the regulatory process.

Proteomics

Characterizing the protein components of a biologic system and elucidating their functions are key factors in understanding the toxicity that may result from biochemical-pathway disruptions or malfunctions due to environmental exposures. Proteomic technologies, such as two-dimensional gel electrophoresis and mass spectrometry, provide avenues for measuring protein-expression changes in response to exposure, identifying the proteins, and characterizing protein modifications, function, and activity (Bandara and Kennedy 2002; Kennedy 2002). Microarray technologies can also be applied to the study of proteins, but they are still in the early development stages.

Many of the proteome investigations address issues in toxicology. The most common form of analysis is differential proteome profiling, which determines the relative expression levels of proteins within a system and may also give information on secondary modifications, such as phosphorylation. The following are examples of proteome profiling experiments:

- Studies that identify protein patterns associated with toxicity, such as acetaminophen-induced toxicity (Fountoulakis et al. 2000), azoxymethane-induced colon tumors (Chaurand et al. 2001), cardiotoxicity (Petricoin et al. 2004), and drug-induced steatosis in liver (Meneses-Lorente et al. 2004). Those studies have also examined dose-response relationships.

- Studies that identify protein biomarkers of effect, such as biomarkers of liver toxicity (Gao et al. 2004) and biomarkers of compound-induced skeletal muscle toxicity (Dare et al. 2002).

- Studies that provide insights into toxicity mechanisms, such as those of biliary canalicular membrane injury (Jones et al. 2003).

- Studies that investigate species differences by proteome characterization of organs and organelles, such as liver proteins in rats (Fountoulakis and Suter 2002) and proteins in liver mitochondrial inner membranes in mice (Da Cruz et al. 2003).

Proteome characterization—determination of the composition of the proteins in a specific system—is a first step in understanding mechanisms of action and the biochemical processes behind induced toxicities. Characterizing the protein differences between species may assist in understanding the differences in species' responses to toxicants. Other proteomic analyses include profiling of protein isoforms and modifications, investigation of protein-protein interactions, and characterization of protein-binding sites related to toxic events (Leonoudakis et al. 2004; Nisar et al. 2004).

Major challenges in proteomics include determining the most appropriate technology to use, processing and interpreting the experimental data, and placing the findings in the correct biologic context. New technologies for differential-expression analysis continue to emerge rapidly, and many are in the validation phase (Zhu et al. 2003). However, difficulties arise when one tries to compare datasets that have been acquired with different technologies, compiled at different times, or generated from different laboratories. Those variations can produce datasets that may or may not lead to the same conclusion (Baggerly et al. 2004). Integrating other types of experimental data, such as genomics datasets, provides additional value and may aid in interpretation (Heijne et al. 2003; Ruepp et al. 2002).

Characterizing various proteomes and using the findings to identify and understand toxicologic events is an enormous undertaking. The Human Proteome Organization (HUPO) was formed in 2001 and has members in various government, industry, and academic organizations (Omenn 2004). One of HUPO's goals is to compare the various technologies that can be used to profile proteomes. There are also plans to develop a comprehensive characterization of the proteins found in human serum and plasma, to evaluate differences within the human population, and to create a global knowledge base and data repository. Concerted efforts, such as HUPO, will expedite our understanding of the proteome, and similar efforts will be needed to answer toxicity-related questions.

Metabonomics

Metabonomics is defined as the study of metabolic responses to drugs, environmental agents, and diseases and involves the quantitative measurement of changes in metabolites in living systems in response to internal or external stimuli or as a consequence of genetic change (Nicholson et al. 2002). The term is often used interchangeably with metabolomics, which is related more specifically to the analysis of *all* metabolites in a biologic sample. Metabonomics is a logical extension of the more established fields of genetics, genomics, and proteomics and is increasingly used as a research tool to characterize chemical-induced changes in physiological processes.

The technique normally involves the processing of biologic fluids—such as urine, plasma, and cerebral spinal fluid—or other tissue preparations followed by analysis with high-resolution nuclear magnetic resonance spectra to identify the metabolites present. As in genomics and proteomics, a large amount of data is generated, and sophisticated computational methods are needed to reduce the "noise" and identify the important changes (Forster et al. 2002). Combining data from multiple -omics sources can give a more holistic understanding of mechanistic toxicology. Mechanistic understanding of even relatively well-characterized agents can be increased by such a combinatorial approach, as recently demonstrated in studies on acetaminophen, which have characterized both genomic and metabonomic end points (Coen et al. 2004).

Many researchers believe that metabonomics can be used in the commercial sector to characterize potential adverse drug effects (Nicolson et al. 2002; Robosky et al. 2002) and as a complementary approach to other -omics technologies in toxicology research (Reo 2002). The pharmaceutical sector has partnered with academia in the Consortium for Metabonomic Toxicology to define appropriate methods and to generate metabolic "fingerprints" of potential use in preclinical screening of candidate drugs (Lindon et al. 2004). Metabonomics may also be used to characterize the effect of environmental stressors in wildlife populations by identifying and characterizing metabolic biomarkers as an indication of organism health. For example, metabonomics has been used to study the "withering syndrome" in shellfish (Viant et al. 2003). The risks and benefits related to novel or engineered food products or mixtures, such as "nutriceuticals," may also be clarified by metabolic assessment of possible consumers or appropriate model systems (German et al. 2003).

Computational Toxicology

Computational toxicology, as defined by EPA (EPA 2003), is the application of mathematical and computer models to predict the effect of an environmental agent and elucidate the cascade of events that result in an adverse response. It uses technologies developed in computational chemistry (computer-assisted simulation of molecular systems), molecular biology (characterization of genetics, protein synthesis, and molecular events involved in biologic response to an agent), bioinformatics (computer-assisted collection, organization, and analysis of large datasets of biologic information), and systems biology (mathematical modeling of biologic systems and phenomena). The goals of using computational toxicology are to set priorities among chemicals on the basis of screening and testing data and to develop predictive models for quantitative risk assessment. Computational toxicology, like the other nonanimal approaches to toxicology previously discussed, holds the potential to lessen the tension between the four major objectives of regulatory testing schemes—breadth, depth, animal welfare, and conservation. Although computational-modeling approaches have the clear advantages of being rapid and of potentially reducing animal testing, the success and validation of any given computational approach clearly depend on the end point being modeled—is the end point amenable to a computational approach?—and on the quality, volume, and chemical diversity contained in the dataset used to generate the model. Some of the computational toxicology products available today are proprietary. To be valuable for risk assessment, the computational approaches must be validated, adequately explained, and made accessible to peer review. Models that are not accessible for review may be useful for many scientific purposes but are not appropriate for regulatory use.

This section first discusses several well-defined modeling activities that have emerged, including structure-activity-relationship (SAR) models, physiologically based pharmacokinetic (PBPK) models, and biologically based dose-response (BBDR) models. It then discusses emerging computational modeling activities, including computational models that predict metabolic fate and three-dimensional models that predict protein-ligand interactions. Finally, it discusses the integration of the various technologies.

Structure-Activity Relationships

The fundamental premise of SAR analyses is that molecular structure determines chemical and physical properties, which determine biologic and toxicologic responses. SAR analyses attempt to answer the questions, What constitutes a class of molecules that are active? What determines relative activity? What distinguishes these from inactive classes? (McKinney et al. 2000). The analyses can be qualitative or quantitative. Generally, SAR analyses are qualitative analyses that predict biologic activity on an ordinal or categorical scale, whereas quantitative SAR (QSAR) analyses use statistical methods to correlate structural descriptors with biologic responses and predict biologic activity on an interval or continuous scale.

SAR and QSAR techniques have been applied to a wide variety of toxicologic end points. They have been used to predict LD_{50} values, maximum tolerated doses, *Salmonella typhimurium* (Ames) assay results, carcinogenic potential, and developmental-toxicity effects. SAR approaches also are used by EPA to screen new industrial chemicals under the Toxic Substances Control Act (TSCA) program. However, some toxicologic end points—such as carcinogenicity, reproductive effects, and hepatotoxicity—are mechanistically ill-defined at the molecular level and this leads to added complexity when one tries to build predictive models for these end points.

Numerous SAR and QSAR modeling packages are commercially available. They are in two main categories: knowledge-based approaches and statistically based systems. Knowledge-based systems, such as DEREK, use rules about generalized relationships between structure and biologic activity that are derived from human expert opinion and interpretation of toxicologic data to predict the potential toxicity of novel chemicals (LHASA Ltd. 2005a). Statistically based systems use calculated measures, structural connectivity, and various statistical methods to derive mathematical relationships for a training set of noncongeneric compounds. Examples of the latter approach are MultiCASE (Multi-CASE 2005) and MDL QSAR (Elsevier MDL 2005).

Physiologically Based Pharmacokinetic Models

PBPK models predict distribution of chemicals throughout the body and describe the interactions of chemicals with biologic targets. For example, PBPK models might help to predict rates of appearance of metabolites or reaction products in target tissues or cellular consequences of interactions, such as mutation or impaired proliferation. PBPK models offer great promise for extrapolating tissue doses and responses from high dose to low dose and for extrapolating from test animals to humans and from one exposure route to another. Over the past 25 years, these models have been developed for a broad array of environmental compounds and drugs and have found diverse applications in reducing uncertainties in chemical risk assessments (Reddy et al. 2005). Specifically, PBPK models have helped to define areas of uncertainty and variability in risk assessment and to show explicitly how variability and uncertainty influence toxicity testing and data interpretation. A variety of software tools are now available to support PBPK modeling, including analytic approaches for sensitivity and variability analyses and for Markov-chain Monte Carlo optimization techniques. Some have the expectation today that PBPK models should be available for dose-response assessment and exposure analysis of all important environmental chemicals.

Another use of PBPK models is human exposure surveillance monitoring. Concentrations of a variety of exogenous chemicals in human tissues, blood, and excreta are often measured (CDC 2005). PBPK models can be used in a form of reverse dosimetry to reconstruct the intensity of exposure required to give specific concentrations in tissues, blood, or excreta of exposed humans. The combination of PBPK analysis with biomonitoring results promises to provide improved measurement of human exposure, which can lead to more precise estimates of risks in exposed human populations.

Biologically Based Dose-Response Models

PBPK models can describe the relationship of dose with an initial biologic response, but BBDR models describe the progression from the initial biologic response through the biologic events leading to alteration of tissue function and disease. They predict the dose-response relationship on the basis of principles of biology, pharmacokinetics, and chemistry. Development of BBDR models requires collection of specific mechanistic data and their organization through quantitative, iterative modeling of biologic processes. The datasets involved in BBDR model construction, particularly those for toxicologic evaluations, will rely increasingly on high-throughput, high-content genomic data to assess cell signaling pathways perturbed by exposure to chemicals and the concentration at which the perturbations become large enough to alter specific biologic processes.

One use emphasized in EPA's computational-toxicology framework is characterizing pathways of toxicity (EPA 2003). The key aspect is identifying the initial biologic alteration that leads to the observed adverse effect. For example, a group of structurally similar chemicals may interact with a specific nuclear receptor and cause a cascade of events, which may be species-specific or tissue-specific but lead to a similar adverse response. Identifying the initial biologic interaction that precipitates the observed adverse effect creates a foundation on which to develop generic BBDR models, that is, BBDR models for classes of compounds.

The main goal in developing BBDR models is to use the validated models to refine low-dose and interspecies extrapolation. Such application would require careful analysis of variability, sensitivity, and robustness of various model structures. BBDR models also may be used to improve the experimental design of toxicology studies so that data needs for risk assessment are fulfilled.

Computational Approaches to Predicting Metabolic Fate

Predicting metabolic fate is important in determining the risks associated with environmental exposure to chemicals. In some cases, for example, the parent compounds are benign but are metabolized to reactive intermediates that form protein or DNA adducts that elicit a toxicologic response. Therefore, identifying the potential metabolites and likely clearance routes is critical for a complete hazard and risk assessment.

Numerous metabolic-fate computational models have been reported, and several are commercially available. Examples of products are METEOR (LHASA Ltd. 2005b), Meta (MultiCASE Inc. 2005), MetaDrug (GeneGo Inc. 2005), and, more recently, MetaSite (Molecular Discovery Ltd. 2005). METEOR, Meta, and MetaDrug use a rule-based approach to biotransformation in that they recognize structural motifs that are susceptible to metabolism and use "weighting" algorithms to determine the most likely metabolic products. Those systems have focused on a general mammalian model but in some cases have been able to generate species-specific predictions where knowledge is available. MetaDrug and MetaSite can predict the most likely sites of metabolism and the responsible P450 enzyme. The predictions rely on three-dimensional models of the individual cytochrome active sites. However, the products are based largely on three-dimensional structure models that have been extrapolated from crystallography data from microbial or other nonhuman P450 enzymes and suffer from the limitations of three-dimensional modeling described below.

Three-Dimensional Modeling of Chemical-Target Interactions

Three-dimensional modeling of a protein-ligand interaction raises the possibility of assessing structures on the basis of a computed ligand docking score (see Jones et al. 1997; Abagayan and Totrov 2001). Many docking software products and scoring algorithms have been developed and are commercially available through organizations, such as the Cambridge Crystallographic Data Center (CCDC 2005; Tripos 2005; Accelrys 2005; Schrödinger 2005). Those methods often assume a flexible ligand but a rigid binding site; they also assume that a single binding site is responsible for the inhibition or activation of the protein function. Such assumptions may not hold true for "promiscuous" receptors, such as the estrogen receptor, that have broad substrate specificity and multiple potential binding ligands. Recent developments in software and advances in computing power have enabled some companies to develop potential solutions to the difficult and computationally expensive problem of dealing with those receptors.

The major limitation in hazard prediction is a general lack of knowledge about protein-ligand interactions mechanistically involved in observed toxicity. Accordingly, most of the focus has been placed on proteins that have been identified as potentially important in toxicology, namely the P450 family of cytochromes, which is thought to be primarily responsible for most drug metabolism,[4] and the human ether-a-go-go (hERG) potassium channel, which is thought to play a role in cardiac QT prolongation considered by several regulatory bodies, including FDA, to be a surrogate indicator of potential drug-induced cardiac arrhythmia.[5]

No x-ray crystal structures for the human variants of the cytochrome P450 proteins are in the public domain. Therefore, most efforts have focused on homology models constructed from bacterial or other mammalian protein structures. However, some commercial vendors claim to have human x-ray structures available for use with drug-design models (Astex Technology 2005). Successful use of three-dimensional modeling techniques, other than that discussed above, has not been widely published. The value of these approaches in predictive toxicology remains to be determined.

Future Uses of Computational Toxicology

Clearly, the computational approaches discussed here represent a set of related scientific disciplines that continue to mature. Their placement in a testing strategy will depend on what questions they can address. For example, regulatory scientists, such as Richard (1998), have commented that the opportunities offered by SAR approaches are most likely to be in hazard identification; this reflects the current inability of these systems (or any other system) to rule out hazard definitively. Nevertheless, it seems clear that these evolving computational tools have an opportunity to contribute substantially to the early stages of a more holistic toxicity-testing strategy. Inevitably, the value of a specific approach will be somewhat context-sensitive and will depend on the robustness of the model and on the quality of the underlying data that support it.

There are opportunities to link PBPK models with SAR approaches. PBPK models require a variety of input parameters, including partition coefficients, metabolic parameters, and rates of metabolism of test compounds. A long-term goal has been to use SAR approaches to provide those input parameters and produce generic PBPK models for

---

[4]For examples of computational modeling of P450s, see Ekins et al. 2001; de Groot et al. 1999; Payne et al. 1999.

[5]For computational-modeling examples, see Aronov and Goldman 2004.

classes of chemicals that vary quantitatively with their structures and with the associated inputs. Continuing improvements in computational methods, especially with regard to predicting metabolic rates and sites of metabolism on complex molecules, could make the technologies feasible in the relatively near future. It may eventually be possible to link SAR, PBPK, and BBDR models to predict dose-response behaviors for the perturbations of cellular signaling networks by exogenous compounds and to provide estimates of risk to exposed humans.

VALIDATION

New or revised toxicity-testing methods for regulatory toxicology are developed for a number of reasons, such as to increase chemical throughput, to provide more detailed information about individual chemicals, to reduce animal use and suffering, and to decrease costs associated with testing. A new or revised method may satisfy one or more of those objectives, indicating that they are not necessarily in conflict. Regardless of the rationale for developing a new method, any such method should be evaluated objectively—that is, validated—to determine whether it fulfills its intended purpose.

The need for formal principles of validation in toxicity testing became evident in the middle to late 1980s when various in vitro tests were developed as potential alternatives to established in vivo tests. The question arose as to how the new tests should be objectively assessed to determine whether they were as good as or better than the existing animal tests in predicting toxicity. Scientists and regulators recognized that formal validation principles would facilitate the implementation of new testing methods that could replace, reduce, or refine animal use and of any methods that involved new and improved technologies or helped to address new regulatory needs. Such validation principles would also help to ensure that the assessment of new methods was conducted in a scientifically sound and high-quality manner.

As a result of several workshop reports that discussed the conceptual and practical aspects of validation (Balls et al. 1990, 1995; OECD 1996), key terms were defined. Validation of a test method is a process by which the reliability and relevance of the method for a specific purpose are established (Balls et al. 1990). The reliability of a test method is the extent of reproducibility of results within and between laboratories over time when the test is performed using the same protocol. Relevance is related to the scientific basis of the test system and to its predictive capability. Predictions are sometimes made with the help of a model, which translates the results from the test system into a prediction of toxicity. Test methods should be both reliable and relevant, and their limitations duly noted. Because in vivo mammalian models are currently assumed to have some relevance to humans, they are generally used as the standard against which alternative models are validated. In the validation of new in vivo mammalian assays, there has been some confusion about how relevance should be assessed. In such cases, validation is directed primarily at determining reproducibility, although relevance remains an important consideration.

Validation is one of several phases in the evolution of a test method from conception to application. The phases are test development, prevalidation or test optimization, formal validation, independent assessment, and regulatory acceptance. Validation is often a time-consuming and expensive process. Consequently, a prevalidation or optimization phase is considered necessary to ensure that a method is ready to enter the validation process. Prevalidation addresses, at least in a preliminary way, many of the issues addressed later in the validation phase, especially the availability of an optimized protocol.

In the validation of new in vivo mammalian bioassays, the adequacy of the test method's end points to evaluate the biologic effect of interest in the species of interest may be difficult to determine. Ideally, the results of the in vivo mammalian bioassay should be compared with results of human studies. However, it is often difficult to validate a mammalian bioassay against health effects in humans because of the lack of high-quality data in humans and ethical constraints in conducting human clinical studies. Therefore, the validation principles discussed below are more easily applied to the validation of nonmammalian assays.

ECVAM was established in 1992 to coordinate validation activities in the European Union, and its U.S. counterpart, ICCVAM, was established in 1994. ECVAM has been more active in coordinating validation exercises, whereas ICCVAM has focused more on assessing the validation status of methods submitted for consideration. Both ECVAM and ICCVAM follow the validation principles developed at an OECD workshop in 1996 (OECD 1996). The principles are intended to apply to the validation of new or updated in vivo or in vitro test methods for hazard assessment. They address such issues as the scientific and regulatory rationale for the proposed test method, the adequacy of the test method's end points to evaluate the biologic effect of interest in the species of interest, the availability of a detailed, formal protocol for the test method, the reproducibility of the test within and among laboratories, the performance of the test method relative to the performance of the test it is designed to replace (if appropriate), the availability of the supporting data for review, and the adherence to good laboratory practices. The international consistency with the validation process is important because validation studies, which can be resource-intensive, expensive, and time-consuming, do not need to be repeated simply to satisfy differing international requirements.

ICCVAM and ECVAM have been increasingly collaborative on projects to improve their collective efficiency. However, they face several challenges in validating toxicologic methods, including those that can replace, reduce, and refine existing animal-based tests. The challenges are as follows:

• The expense, time, and resources entailed by some validation efforts are impediments to more rapid progress. ICCVAM should strive for ways to overcome the logistical constraints without compromising the scientific integrity of the process. ECVAM's "modular" approach to validation may be helpful in this regard. The modular approach decouples the stepwise process and emphasizes the data needed to address various principles of validation, such as within- and between-laboratory variability. The data needs are regarded as discrete modules, each of which can be satisfied with a distinct set of data, some of which may be derived from pre-existing data. Although new data may be needed, the number of laboratories required may be smaller than in a standard validation exercise (Hartung et al. 2004).

• Many validation efforts compare a new test method undergoing validation with an existing animal-based test for the same end point. Such comparisons necessarily require comprehensive data not only on the new method but also on the existing method—the reference test. Experience has shown that the data from such reference tests are limited in availability. Test results, if published at all, are often provided in summary form, whereas ICCVAM typically needs individual animal data. The challenge to ICCVAM is to work with industry and others to assemble as complete a set of high-quality animal data as possible. Such efforts, when successful, would preclude the need to conduct further animal testing to generate new data.

• Ideally, human data should serve as the standard against which to evaluate the performance of new tests. In the absence of such human data, ICCVAM and similar entities consider the existing test to be the default standard and judge the new test against it. A challenge arises when the reference tests, typically animal-based, have considerable variability across laboratories. Such variability makes it difficult to show correlations between the results of the new test method and those of the reference test. One way to address this challenge is to make greater efforts to collect available human data as the true standard for comparison. In the absence of such data, however, approaches need to be developed to account for the inherent variability in some animal tests when conducting validation assessments.

- New test methods are not always stand-alone substitutes for existing test methods. New test methods that prove to be inadequate in head-to-head comparisons with existing test methods might pass muster when combined with complementary approaches into tiered or battery approaches. Consequently, ICCVAM might benefit from providing greater guidance on developing and validating such approaches, rather than relying on one-for-one correspondence between the new and existing test methods.

- Another challenge facing ICCVAM is helping to ensure a steady flow of new test methods into its validation pipeline. Without such candidate methods, ICCVAM would have nothing to validate or assess. ICCVAM or its parent agency should consider funding research to identify biomarkers or mechanisms of toxicity that could be incorporated into test methods and channeled into the ICCVAM pipeline for validation.

Meeting the challenges discussed above would enable ICCVAM to be more productive and efficient in assessing new test methods for their suitability for regulatory toxicology.

In addition to its guidance on validation principles, ICCVAM and the NTP Interagency Center for the Evaluation of Alternative Toxicological Methods (NICEATM) have issued practical guidance on submitting validation data for assessment and nominating promising test methods for further development or validation (ICCVAM/NICEATM 2004). Several new or revised tests have gone through the ICCVAM process and have been assessed according to its validation and regulatory acceptance criteria. For example, in 1998, after a submission by industry representatives, ICCVAM established an independent peer-review panel to review the validation status of the local lymph node assay (LLNA), a reduction and refinement alternative to the guinea pig maximization test

(GPMT) test for allergic contact dermatitis. The panel judged the LLNA to be an adequate substitute for the GPMT according to the ICCVAM validation criteria. ICCVAM forwarded the results of the review to relevant federal agencies, which accepted the LLNA as a validated test for allergic contact dermatitis.

The ICCVAM-NICEATM validation and submission criteria are intended to help industry and the federal government to update and enhance the inventory of chemical testing methods. New or revised methods can be reviewed by ICCVAM and NICEATM, and the resulting recommendations can be sent to individual agencies for their consideration. Thus, the guidelines can help stakeholders to meet the challenges posed by new testing programs or needs. For example, EPA has contracted with ICCVAM and NICEATM to validate receptor-binding assays for its endocrine-disruptor program, and it is using ICCVAM and NICEATM criteria to validate some animal-based tests for the program. It should be emphasized that the formal validation process applies to methods intended for immediate regulatory testing. It is not intended for methods that, for example, are used only inhouse in industry or are purely investigational or newly emerging.

REFERENCES

Abagyan, R., and M. Totrov. 2001. High-throughput docking for lead generation. Curr. Opin. Chem. Biol. 5(4):375-382.

Accelrys. 2005. Products and Services. Accelrys Software Inc. [online]. Available: http://www.accelrys.com [accessed April 12, 2005].

Andreasen, E.A., J.M. Spitsbergen, R.L. Tanguay, J.J. Stegeman, W. Heideman, and R.E. Peterson. 2002. Tissue-specific expression of AHR2, ARNT2, and CYP1A in zebrafish embryos and larvae: Effects of developmental stage and 2,3,7,8- tetrachlorodibenzo-*p*-dioxin exposure. Toxicol. Sci. 68 (2):403-419.

Aronov, A.M., and B.B. Goldman. 2004. A model for identifying HERG K+ channel blockers. Bioorg. Med. Chem. 12(9):2307-2315.

Astex Technology. 2005. Current Portfolio. Astex Technology, Cambridge, UK [online]. Available: http://www.astex-technology.com/current_ portfolio.html [accessed April 12, 2005].

Baggerly, K.A., J.S. Morris, and K.R. Coombes. 2004. Reproducibility of SELDI-TOF protein patterns in serum: Comparing datasets from different experiments. Bioinformatics 20(5):777-785.

Balls, M., P. Botham, A. Cordier, S. Fumero, D. Kayser, H. Koëter, P. Koundakjian, N.G. Lindquist, O. Meyer, L. Pioda, C. Reinhardt, H. Rozemond,

T. Smyrniotis, H. Spielmann, H. Van Looy, M.T. van der Venne, and E. Walum. 1990. Report and recommendations of an international workshop on promotion of the regulatory acceptance of validated non-animal toxicity test procedures. ATLA 18:339-344.

Balls, M., A.N. Goldberg, J.H. Fentem, C.L. Broadhead, R.L. Burch, M.F.W. Festing, J.M. Frazier, C.F.M. Hendricksen, M. Jennings, M.D.O. van der Kamp, D.B. Morton, A.N. Rowan, C. Russel, W.M.S. Russell, H. Spielmann, M.L. Stephens, W.S. Stokes, D.W. Straughan, J.D. Yager, J. Zurlo, and B.F.M. van Zutphen. 1995. The three Rs: The way forward. ATLA 23(6):838-866.

Bandara, L., and S. Kennedy. 2002. Toxicoproteomics—a new preclinical tool. Drug Discov. Today 7(7):411-418.

Berry, D.A., P. Mueller, A.P. Grieve, M. Smith, T. Parke, R. Balazek, N. Mitchard, and M. Krams. 2002. Adaptive Bayesian designs for dose-ranging drug trials. Pp. 99-181 in Case Studies in Bayesian Statistics, Vol. V, C. Gatsonis, R.E. Kass, B. Carlin, A. Carriquiry, A. Gelman, I. Verdinelli, and M. West, eds. New York, NY: Springer.

CCDC (Cambridge Crystallographic Data Centre). 2005. Products. Cambridge Crystallographic Data Centre, Cambridge, UK [online]. Available: http:// www.ccdc.cam.ac.uk [accessed April 12, 2005].

CDC (Centers for Disease Control and Prevention). 2005. Third National Report on Human Exposure to Environmental Chemicals. U.S. Department of Health and Human Services, Centers for Disease Control and Prevention, Atlanta, GA [online]. Available: http://www.cdc.gov/ exposurereport/3rd/pdf/thirdreport.pdf [accessed Sept. 26, 2005].

Chaurand, P., B.B. DaGue, R.S. Pearsall, D.W. Threadgill, and R.M. Caprioli. 2001. Profiling proteins from azoxymethane-induced colon tumors at the molecular level by matrix-assisted laser desorption/ionization mass spectrometry. Proteomics 1(10):1320-1326.

Coen, M., S.U. Ruepp, J.C. Lindon, J.K. Nicholson, F. Pognan, E.M. Lenz, and I.D. Wilson. 2004. Integrated application of transcriptomics and metabonomics yields new insight into the toxicity due to paracetamol in the mouse. J. Pharm. Biomed. Anal. 35(1):93-105.

Contag, C.H., P.R. Contag, S.D. Spilman, D.K. Stevenson, and D.A. Benaron. 1996. Photonic monitoring of infectious disease and gene regulation. Pp. 220-224 in Biomedical Optical Spectroscopy and Diagnostics, E. Sevick-Muraca, and D. Benaron, eds. Trends in Optics and Photonics Vol. 3. Washington, DC: Optical Society of America.

Daborn, P.J., J.L. Yen, M.R. Bogwitz, G. Le Goff, E. Feil, S. Jeffers, N. Tijet, T. Perry, D. Heckel, P. Batterham, R. Feyereisen, T.G. Wilson, and R.H. ffrench-Constant. 2002. A single p450 allele associated with insecticide resistance in *Drosophila*. Science 297(5590):2253-2256.

Da Cruz, S., I. Xenarios, J. Langridge, F. Vilbois, P.A. Parone, and J.C. Martinou. 2003. Proteomic analysis of the mouse liver mitochondrial inner membrane. J. Biol. Chem. 278(42):41566-41571.

Dare, T., H.A. Davies, J.A. Turton, L. Lomas, T.C. Williams, and M.J. York. 2002. Application of surface-enhanced laser desorption/ionization technology to the detection and identification of urinar parvalbumin-alpha: A biomarker of compound-induced skeletal muscle toxicity in the rat. Electrophoresis 23(18):3241-3251.

de Groot, M.J., M.J. Ackland, V.A. Horne, A.A. Alex, and B.C. Jones. 1999. A novel approach to predicting P450 mediated drug metabolism. CYP2D6 catalyzed N-dealkylation reactions and qualitative metabolite predictions using a combined protein and pharmacophore model for CYP2D6. J. Med. Chem. 42(20):4062-4070.

Diehl, K.H., R. Hull, D. Morton, R. Pfister, Y. Rabemampianina, D. Smith, J.M. Vidal, and C. van de Vorstenbosch. 2001. A good practice guide to the administration of substances and removal of blood, including routes and volumes. J. Appl. Toxicol. 21(1):15-23.

Dolan, M.E., K.G. Newland, R. Nagasubramanian, X. Wu, M.J. Ratain, E.H. Cook Jr., and J.A. Badner. 2004. Heritability and linkage analysis of sensitivity to cisplatin-induced cytotoxicity. Cancer Res. 64(12):4353-4356.

EBI (European Bioinformatics Institute). 2005. Microarray, Tox-MIAMExpress. European Bioinformatics Institute, European Molecular Biology Laboratory [online]. Available: http://www.ebi.ac.uk/tox-miamexpress/ [accessed April 12, 2005].

ECVAM (European Centre for the Validation Alternative Methods). 2005. Scientifically Validated Methods [online]. Available: http://ecvam.jrc. cec.eu.int/index.htm [accessed March 16, 2005].

Ekins, S., M.J. de Groot, and J.P. Jones. 2001. Pharmacophore and three-dimensional quantitative structure activity relationship methods for modling cytochrome p450 active sites. Drug Metab. Dispos. 29(7):936-944.

Elsevier MDL. 2005. MDL QSAR. MDL Discovery Predictive Science [online]. Available: http://www.mdl.com/products/predictive/qsar/index. jsp [accessed April 12, 2005].

EPA (U.S. Environmental Protection Agency). 1998. Health Effects Test Guidelines OPPTS 870.2400. Acute Eye Irritation. EPA 712-C-98-195. Office of Prevention, Pesticides, and Toxic Substances, U.S. Environmental Protection Agency, Washington, DC [online]. Available: http://www.epa.gov/opptsfrs/OPPTS_Harmonized/870_Health_Effects_Te st_Guidelines/Series/870-2400.pdf [accessed April 7, 2005]

EPA (U.S. Environmental Protection Agency). 1999. Letters to Manufacturers/Importers, October 14, 1999. High Production Volume Challenge Program, Office of Prevention, Pesticides, and Toxic Substances, U.S. Environmental Protection Agency, Washington, DC [online]. Available: http://www.epa.gov/chemrtk/ceoltr2.htm [accessed April 11, 2005].

EPA (U.S. Environmental Protection Agency). 2002. Interim Policy on Genomics. Science Policy Council, Office of the Science Advisor, U. S. Environmental Protection Agency, Washington, DC [online]. Available: http://epa.gov/osa/spc/htm/genomics.htm [accessed April 12, 2005].

EPA (U.S. Environmental Protection Agency). 2003. A Framework for a Computational Toxicology Research Program in ORD. Draft Report. EPA/600/R-03/065. Office of Research and Development, U.S. Environmental Protection Agency. July 2003 [online]. Available: http://www.epa.gov/sab/03minutes/ctfcpanel_091203mattach_e.pdf [accessed April 7, 2005].

EPA (U.S. Environmental Protection Agency). 2004. Potential Implications of Genomics for Regulatory and Risk Assessment Applications at EPA. External Review Draft. EPA 100/B-04/002. Genomics Task Force Workgroup, Science Policy Council, U. S. Environmental Protection Agency, Washington, DC [online]. Available: http://www.epa.gov/osa/ genomics-external-review-draft.pdf [accessed April 12, 2005].

FDA (Food and Drug Administration). 2005. Guidance for Industry: Pharmacogenomic Data Submissions. Center for Drug Evaluation and Research, Center for Biologics Evaluation and Research, Center for Devices and Radiological Health, Food and Drug Administration. March 2005 [online]. Available: http://www.fda.gov/cder/guidance/6400fnl.htm [accessed June 3, 2005].

Festing, M.F.W. 1999. Reduction in animal use in the production and testing of biologicals. Dev. Biol. Stand. 101:195-200.

Festing, M.F.W., V. Baumans, R.D. Combes, M. Halder, C.F.M. Hendriksen, B.R. Howard, D.P. Lovell, G.J. Moore, P. Overend, and M.S. Wilson. 1998. Reducing the use of laboratory animals in biomedical research: Problems and possible solutions. ATLA 26(3):283-301 [online]. Available: http://altweb.jhsph.edu/publications/ECVAM/ecvam29.htm [accessed April 7, 2005].

Forster, J., A.K. Gombert, and J. Nielsen. 2002. A functional genomics approach using metabolomics and in silico pathway analysis. Biotechnol. Bioeng. 79(7):703-712.

Fountoulakis, M., and L. Suter. 2002. Proteomic analysis of the rat liver. J. Chromatogr. B Analyt. Technol. Biomed. Life Sci. 782(1-2):197-218.

Fountoulakis, M., P. Berndt, U.A. Boelsterli, F. Crameri, M. Winter, S. Albertini, and L. Suter. 2000. Two-dimensional database of mouse liver proteins: Changes in hepatic protein levels following treatment with acetaminophen or its nontoxic regioisomer 3-acetamidophenol. Electrophoresis 21(11):2148-2161.

Gao, J., L.A. Garulacan, S.M. Storm, S.A. Hefta, G.J. Opiteck, J.H. Lin, F. Moulin, and D. Dambach. 2004. Identification of in vitro protein biomarkers of idiosyncratic liver toxicity. Toxicol. In Vitro 18(4):533-541.

Gaylor, D.W. 1980. The ED01 study: Summary and conclusions. J. Environ. Pathol. Toxicol. 3(3 Spec.):179-183.

GeneGo Inc. 2005. MetaDrug. GeneGo Inc., St. Joseph, MI [online]. Available: http://www.genego.com/about/products.shtml#metadrug [accessed April 12, 2005].

German, J.B., M.A. Roberts, and S.M. Watkins. 2003. Genomics and metabolomics as markers for the interaction of diet and health: Lessons from lipids. J. Nutr. 133(6 Suppl. 1):2078S-2083S.

Giles, F.J, H.M. Kantarjian, J.E. Cortes, G. Garcia-Manero, S. Verstovsek, S. Faderl, D.A. Thomas, A. Ferrajoli, S. O'Brien, J.K. Wathen, L.C. Xiao, D.A. Berry, and E.H. Estey. 2003. Adaptive randomized study of idarubicin and cytarabine versus troxacitabine and cytarabine versus troxacitabine and idarubicin in untreated patients 50 years or older with adverse karyotype acute myeloid leukemia. J. Clin. Oncol. 21(9):1722-1727.

Greenwood, T.A., P.E. Cadman, M. Stridsberg, S. Nguyen, L. Taupenot, N.J., Schork, and D.T. O'Connor. 2004. Genome-wide linkage analysis of chromogranin B expression in CEPH pedigrees: Implications for exocytotic sympathochromaffin secretion in humans. Physiol Genomics. 18(1):119-127.

Hartung, T., S. Bremer, S. Casati, S. Coecke, R. Corvi, S. Fortaner, L. Gribaldo, M. Halder, S. Hoffmann, A.J. Roi, P. Prieto, E. Sabbioni, L. Scott, A. Worth, and V. Zuang. 2004. A modular approach to the ECVAM principles on test validity. ATLA 32(5):467-472.

Heijne, W., R.H. Stierum, M. Slijper, P.J. van Bladeren, and B. van Ommen. 2003. Toxicogenomics of bromobenzene hepatotoxicity: A combined transcriptomics and proteomics approach. Biochem. Pharmacol. 65(5): 857-875.

Hendricksen, C.F.M., and D.B. Morton, eds. 1999. Humane Endpoints in Animal Experimentation for Biomedical Research: Proceedings of the International Conference, November 22-25, 1998, Zeist, The Netherlands. London: Royal Society of Medicine Press.

Hood, E. 2004. Taking stock of toxicogenomics: Mini-monograph offers overview [comment]. Environ. Health Perspect. 112(4):A231.

ICCVAM (Interagency Coordinating Committee on the Validation of Alternative Methods). 2004. Test Methods Evaluations [online]. Available: http://iccvam.niehs.nih.gov/methods/review.htm [accessed March 16, 2005].

ICCVAM–NICEATM (Interagency Coordinating Committee on the Validation of Alternative Methods- National Toxicology Program Interagency Center for the Evaluation of Alternative Toxicological Methods). 2004. ICCVAM - NICEATM Documents [online]. Available: http://iccvam. niehs.nih.gov/docs/docs.htm [accessed April 12, 2005].

ILAR (Institute for Laboratory Animal Research). 2000. Humane Endpoints for Animals Used in Biomedical Research and Testing. ILAR J. 41(2):59-

123 [online]. Available: http://dels.nas.edu/ilar/jour_online/41_2/41_2.asp [accessed April 7, 2005].

Jones, G., P. Willett, R.C. Glen, A.R. Leach, and R. Taylor. 1997. Development and validation of a genetic algorithm for flexible docking. J. Mol. Biol. 267(3):727-748.

Jones, J.A., L. Kaphalia, M. Treinen-Moslen, and D.C. Leibler. 2003. Proteomic characterization of metabolites, protein adducts, and biliary proteins in rats exposed to 1,1-dichloroethylene or diclofenac. Chem. Res. Toxicol. 16(10):1306-1317.

Kennedy, S. 2002. The role of proteomics in toxicology: Identification of biomarkers of toxicity by protein expression analysis. Biomarkers 7(4): 269-290.

Leonoudakis, D., L.R. Conti, S. Anderson, C.M. Radeke, L.M. McGuire, M.E. Adams, S.C. Froehner, J.R. Yates, and C.A. Vandenberg. 2004. Protein trafficking and anchoring complexes revealed by proteomic analysis of inward rectifier potassium channel (Kir2.x)-associated proteins. J. Biol. Chem. 279(21):22331-22346.

LHASA Ltd. 2005a. DEREK for Windows. LHASA Limited, Department of Chemistry, University of Leeds, Leeds, UK [online]. Available: http://www.chem.leeds.ac.uk/luk/derek/index.html [accessed April 12, 2005].

LHASA Ltd. 2005b. METEOR. LHASA Limited, Department of Chemistry, University of Leeds, Leeds, UK [online]. Available: http://www.chem.leeds.ac.uk/luk/meteor/index.html [accessed April 12, 2005].

Lindon, J.C., E. Holmes, M.E. Bollard, E.G. Stanley, and J.K. Nicholson. 2004. Metabonomics technologies and their applications in physiological monitoring, drug safety assessment and disease diagnosis. Biomarkers 9(1):1-31.

Lo, H.S., Z. Wang, Y. Hu, H.H. Yang, S. Gere, K.H. Buetow, and M.P. Lee. 2003. Allelic variation in gene expression is common in the human genome. Genome Res. 13(8):1855-1862.

McKinney, J.D., A. Richard, C. Waller, M.C. Newman, and F. Gerberick. 2000. The practice of structure activity relationships (SAR) in toxicology. Toxicol. Sci. 56(1):8-17.

Meneses-Lorente, G., P.C. Guest, J. Lawrence, N. Muniappa, M.R. Knowles, H.A. Skynner, K. Salim, I. Cristea, R. Mortishire-Smith, S.J. Gaskell, and A. Watt. 2004. A proteomic investigation of drug-induced steatosis in rat liver. Chem. Res. Toxicol. 17(5):605-612.

Moennikes, O., S. Loeppen, A. Buchmann, P. Andersson, C. Ittrich, L. Poellinger, and M. Schwarz. 2004. A constitutively active dioxin/aryl hydrocarbon receptor promotes hepatocarcinogenesis in mice. Cancer Res. 64(14):4707-4710.

Molecular Discovery Ltd. 2005. MetaSite. Molecular Discovery Ltd [online]. Available: http://www.moldiscovery.com/soft_metasite.php [accessed April 12, 2005].

Morley, M., C.M. Molony, T.M. Weber, J.L Devlin, K.G. Ewens, R.S. Spielman, and V.G. Cheung. 2004. Genetic analysis of genome-wide variation in human gene expression. Nature 430(7001):743-747.

Morton, D.B. 1995. Advances in refinement in animal experimentation over the past 25 years. ATLA 23(6):812-822.

Morton, D.B., M. Jennings, A. Buckwell, R. Ewbank, C. Godfrey, B. Holgate, I. Inglis, R. James, C. Page, I. Sharman, R. Verschoyle, L. Westfall, and A.B. Wison. 2001. Refining procedures for the administration of substances. Lab. Anim. 35(1):1-41.

MultiCASE Inc. 2005. META Program. MultiCASE Inc., Beachwood, OH [online]. Available: http://www.multicase.com/ [accessed April 12, 2005].

Nicholson, J.K., J. Connelly, J.C. Lindon, and E. Holmes. 2002. Metabonomics: A platform for studying drug toxicity and gene function. Nat. Rev. Drug Discov. 1(2):153-161.

Nisar, S., C.S. Lane, A.F. Wilderspin, K.J. Welham, W.J. Griffiths, and L.H. Patterson. 2004. A proteomic approach to the identification of cytochrome P450 isoforms in male and female rat liver by nanoscale liquid chromatography-electrospray ionization-tandem mass spectrometry. Drug Metab. Dispos. 32(4):382-386.

NTP (National Toxicology Program). 1999. The Murine Local Lymph Node Assay: A Test Method for Assessing the Allergic Contact Dermatitis Potential of Chemicals/Compounds. NIH Publication No. 99-4494. National Toxicology Program, Research Triangle Park, NC. February 1999 [online]. Available: http://iccvam.niehs.nih.gov/methods/llnadocs/llnarep. pdf [accessed March 16, 2005].

OECD (Organisation for Economic Cooperation and Development). 1996. Final Report of the OECD Workshop on Harmonization of Validation and Acceptance Criteria for Alternative Toxicological Test Methods. Paris: OECD.

OECD (Organisation for Economic Cooperation and Development). 2000. Guidance Document on the Recognition, Assessment, and Use of Clinical Signs as Humane Endpoints for Experimental Animals Used in Safety Evaluation. ENV/JM/MONO(2000)7. Paris: OECD [online]. Available: http://www.olis.oecd.org/olis/2000doc.nsf/LinkTo/env-jm-mono(2000)7 [accessed April 7, 2005].

OECD (Organization for Economic Cooperation and Development). 2002a. OECD Test Guideline 401 will be deleted: A Major Step in Animal Welfare: OECD Reaches Agreement on the Abolishment of the LD50 Acute Toxicity Test [online]. Available: http://www.oecd.org/document/ 52/0,2340,en_2649_34377_2752116_1_1_1_1,00.html [accessed March 1, 2005].

OECD (Organisation for Economic Cooperation and Development). 2002b. Acute Eye Irritation/Corrosion. Chemicals Testing Guidelines No. 405.

Paris: Organisation for Economic Cooperation and Development. April 24, 2002.

Omenn, G.S. 2004. The Human Proteome Organization Plasma Proteome Project pilot phase: Reference specimens, technology platform compar-isons, and standardized data submissions and analyses. Proteomics 4(5): 1235-1240.

Payne, V.A., Y.T. Chang, and G.H. Loew. 1999. Homology modeling and substrate binding study of human CYP2C9 enzyme. Proteins 37(2):176-190.

Pedra, J.H., L.M. McIntyre, M.E. Scharf, and B.R. Pittendrigh. 2004. Genome-wide transcription profile of field- and laboratory-selected dichloro-diphenyltrichloroethane (DDT)-resistant Drosophila. Proc. Natl. Acad. Sci. U.S.A. 101(18):7034-7039.

Pennie, W., S.D. Pettit, and P.G. Lord. 2004. Toxicogenomics in risk assessment: An overview of an HESI collaborative research program. Environ. Health Perspect. 112(4):417-419.

Petricoin, E.F., V. Rajapaske, E.H. Herman, A.M. Arekani, S. Ross, D. Johann, A. Knapton, J. Zhang, B.A. Hitt, T.P. Conrads, T.D. Veenstra, L.A. Liotta, and F.D. Sistare. 2004. Toxicoproteomics: Serum proteomic pattern diagnostics for early detection of drug induced cardiac toxicities and cardio-protection. Toxicol. Pathol. 32(Suppl. 1):122-130.

Rea, M.A., J.P. Gregg, Q. Qin, M.A. Phillips, and R.H. Rice. 2003. Global alteration of gene expression in human keratinocytes by inorganic arsenic. Carcinogenesis 24(4):747-756.

Reddy, M.B., R.S.H. Yang, H.J. Clewell III, and M.E. Andersen, eds. 2005. Physiologically Based Pharmacokinetics: Science and Applications. Hoboken, NJ: John Wiley & Sons, Inc.

Reo, N.V. 2002. NMR-based metabolomics. Drug Chem. Toxicol. 25(4):375-382.

Richard, A.M. 1998. Commercial toxicology prediction systems: A regulatory perspective. Toxicol. Lett. (102-103):611-616.

Robosky, L.C., D.G. Robertson, J.D. Baker, S. Rane, and M.D. Reily. 2002. In vivo toxicity screening programs using metabonomics. Comb. Chem. High Throughput Screen. 5(8):651-662.

Ruepp, S., R.P. Tonge, J. Shaw, N. Wallis, and F. Pognan. 2002. Genomics and proteomics analysis of acetaminophen toxicity in mouse liver. Toxicol. Sci. 65(1):135-150.

Russell, W.M.S., and R.L Burch. 1992. The Principles of Humane Experimental Technique, Special Ed. Herts, England: Universities Fed-eration for Animal Welfare. 238 pp.

Schadt, E.E., S.A. Monks, T.A. Drake, A.J. Lusis, N. Che, V. Colinayo, T.G. Ruff, S.B. Milligan, J.R. Lamb, G. Cavet, P.S. Linsley, M. Mao, R.B. Stoughton, and S.H. Friend. 2003. Genetics of gene expression surveyed in maize, mouse and man. Nature 422(6929):297-302.

Schork, N.J., J.P. Gardner, L. Zhang, D. Fallin, B. Thiel, H. Jakubowski, and A. Aviv. 2002. Genomic association/linkage of sodium lithium counter-transport in CEPH pedigrees. Hypertension 40(5):619-628.

Schrödinger. 2005. Product Information. Schrödinger, Portland, OR [online]. Available: http://www.schrodinger.com/index.html [accessed April 12, 2005].

Stephens, M.L. 1989. Replacing animal experiments. Pp. 144-168 in Animal Experimentation: The Consensus Changes, G. Langley, ed. New York: Chapman and Hall.

Stephens, M.L, A.M. Goldberg, and A.N. Rowan. 2001. The first forty years of the alternatives approach: Refining, reducing, and replacing the use of laboratory animals. Pp. 121-135 in The State of the Animals 2001, 1st Ed., D.J. Salem, and A.N. Rowan, eds. Washington, DC: The Humane Society Press.

Stephens, M.L., K. Conlee, G. Alvino, and A.N. Rowan. 2002. Possibilities for refinement and reduction: Future improvements within regulatory testing. ILAR J. 43(Suppl.):S74-S79.

Stokes, W.S. 2000. Introduction: Reducing unrelieved pain and distress in laboratory animals using humane endpoints. ILAR J. 41(2):59-61.

Sumanas, S., and S. Lin. 2004. Zebrafish as a model system for drug target screening and validation [review]. Drug Discov. Today Targets 3(3):89-96.

Terasaka, S., Y. Aita, A. Inoue, S. Hayashi, M. Nishigaki, K. Aoyagi, H. Sasaki, Y. Wada-Kiyama, Y. Sakuma, S. Akaba, J. Tanaka, H. Sone, J. Yonemoto, M. Tanji, and R. Kiyama. 2004. Using a customized DNA microarray for expression profiling of the estrogen-responsive genes to evaluate estrogen activity among natural estrogens and industrial chemicals. Environ. Health Perspect. 112(7):773-781.

Tripos Inc. 2005. Discovery Informatics Products. Tripos Inc., St. Louis, MO [online]. Available: http://www.tripos.com/ [accessed April 12, 2005].

Vaughan, S. 2004. Optimising resources by reduction: The FRAME Reduction Committee. ATLA 32(Suppl. 1):245-248.

Viant, M.R., E.S. Rosenblum, and R.S. Tieerdema. 2003. NMR-based metabolomics: A powerful approach for characterizing the effects of environmental stressors on organism health. Environ. Sci. Technol. 37(21):4982-4989.

Watters, J.W., A. Kraja, M.A. Meucci, M.A. Province, and H.L. McLeod. 2004. Genome-wide discovery of loci influencing chemotherapy cytotoxicity. Proc. Natl. Acad. Sci. U.S.A. 101(32):11809-11814.

Wei, Y.D., K. Tepperman, M.Y. Huang, M.A. Sartor, and A. Puga. 2004. Chromium inhibits transcription from polycyclic aromatic hydrocarbon-inducible promoters by blocking the release of histone deacetylase and preventing the binding of p300 to chromatin. J. Biol. Chem. 279(6):4110-4119.

William, D.E., G.S. Bailey, A. Reddy, J.D. Hendricks, A. Oganesian, G.A. Orner, C.B. Pereira, and J.A. Swenberg. 2003. The rainbow trout (*Oncorhynchus mykiss*) tumor model: Recent applications in low-dose exposures to tumor initiators and promoters. Toxicol. Pathol. 31(Suppl.):58-61.

Worth, A.P., and M. Balls, eds. 2002. Alternative (Nonanimal) Methods for Chemicals Testing: Current Status and Future Prospects. ATLA 30(Suppl. 1).

Yan, H., W. Yuan, V.E. Velculescu, B. Vogelstein, and K.W. Kinzler. 2002. Allelic variation in human gene expression. Science 297(5584):1143.

Zheng, X.H., G.S. Watts, S. Vaught, and A.J. Gandolfi. 2003. Low-level arsenite induced gene expression in HEK293 cells. Toxicology 187(1):39-48.

Zhu, H., M. Bilgin, and M. Snyder. 2003. Proteomics. Annu. Rev. Biochem. 72:783-812.

Appendix A

Biographic Information on the Committee on Toxicity Testing and Assessment of Environmental Agents

Daniel Krewski (*Chair*) is professor of medicine and of epidemiology and community medicine and director of the McLaughlin Centre for Population Health Risk Assessment at the University of Ottawa. Previously, he served as director of risk management and as director of the Bureau of Chemical Hazards with Health Canada. Dr. Krewski is associate editor of Risk Analysis. His research interests include epidemiology, biostatistics, risk assessment, and risk management. He serves on National Research Committee (NRC) Committee on Health Risks from Exposure to Low Levels of Ionizing Radiation (BEIR VII Phase 2). He has previously served on the NRC Board on Environmental Studies and Toxicology, the Board on Radiation Effects Research, the Committee on Toxicology, the Committee on Pesticides in the Diets of Infants and Children, the Subcommittee on Pharmacokinetics, the Committee on Comparative Toxicity of Naturally Occurring Carcinogens, the Subcommittee on Health Assessment of Ingested Fluoride, and the Subcommittee on Acute Exposure Guideline Levels. He recently chaired the NRC's Colloquium on Scientific Advances and the Future in Toxicologic Risk Assessment. Dr. Krewski has published more than 400 journal articles and book chapters on risk assessment, biostatistics, and epidemiology. He received his MSc and PhD in mathematics and statistics from Carleton University and his MHA from the University of Ottawa.

Daniel Acosta, Jr., is dean of the College of Pharmacy at the University of Cincinnati. Dr. Acosta's research focuses on the development of in vitro cellular models to explore and evaluate the mechanisms by which xenobiotics damage cell types. He has worked to develop primary culture systems of liver, heart, kidney, nerve, skin, and eye cells as experimental models to study the cellular and subcellular toxicity of selected xenobiotics. He was elected president of the Society of Toxicology, 2000-2001, and is editor of Toxicology In Vitro and the Target Organ Series on Cardiovascular Toxicology. Dr. Acosta serves as chair of the Food and Drug Administration (FDA) Scientific Advisory Board for the National Center for Toxicology Research and was a member of the Board of Scientific Councilors for the Office of Research and Development for the Environmental Protection Agency (EPA) for 2001-2004. He was a member of the Scientific Advisory Committee to the director of the National Center for Environmental Health of the Centers for Disease Control (CDC) for 2001-2003. He is serving on the Scientific Advisory Committee for Alternative Toxicological Methods for the National Institute for Environmental Health Sciences (NIEHS), and the Expert Committee on Toxicology and Biocompatibility for the US Pharmacopoeia (2000-2005). He also served on the NRC Howard Hughes Medical Institute Predoctoral Fellowships Panel on Neurosciences and Physiology. Dr. Acosta received his PhD in pharmacology and toxicology from the University of Kansas.

Melvin Andersen is director of the Computational Biology Division at the CIIT Centers for Health Research. Previously, he held positions in toxicology research and research management in the federal government (Department of Defense and EPA) and was professor of environmental health at Colorado State University. He has worked to develop biologically realistic models of the uptake, distribution, metabolism, and biologic effects of drugs and toxic chemicals and has applied these physiologically based pharmacokinetic and pharmacodynamic models to safety assessments and quantitative health risk assessments. His current research interests include developing mathematical descriptions of control of genetic circuitry in the developing and adult organism and the dose-response and risk-assessment implications of the control processes. Dr. Andersen is board-certified in industrial hygiene and in toxicology. He has served on numerous NRC committees, including the Committee on Toxicology, the Committee on Toxicological Effects of Mercury, the Committee on Risk Assessment Methodology, and the Subcommittee on Pharmacokinetics. He earned a PhD in biochemistry and molecular biology from Cornell University.

Henry Anderson is chief medical officer and state epidemiologist for occupational and environmental health in the Wisconsin Division of Public Health and adjunct professor of population health at the University of Wisconsin Medical School. Dr. Anderson's research interests include disease surveillance, risk assessment, childhood asthma, lead poisoning, health hazards of Great Lakes sport fish consumption, arsenic in drinking water, bioterrorism, asbestos disease, vermiculite exposure, and occupational fatalities and injuries in youth. He is certified by the American Board of Preventive Medicine with a subspecialty in occupational and environmental medicine and is a fellow of the American College of Epidemiology. Dr. Anderson is chairperson of the Board of Scientific Councilors of the National Institute of Occupational Safety and Health (NIOSH) and has served as chairperson of the Environmental Health Committee of the EPA Scientific Advisory Board. He served on the NRC Committee on a National Agenda for the Prevention of Disabilities and the Committee on Enhancing Environmental Health Content in Nursing Practice. Dr. Anderson received his MD from the University of Wisconsin Medical School.

John C. Bailar III is professor emeritus in the Department of Health Studies at the University of Chicago. He is a retired Commissioned Officer of the US Public Health Service and worked for the National Cancer Institute for 22 years. He has also held academic appointments at Harvard University and McGill University. Dr. Bailar's research interests include assessing health risks posed by chemical hazards and air pollutants and interpreting statistical evidence in medicine, with emphasis on cancer. He was editor-in-chief of the Journal of the National Cancer Institute for 6 years, and was statistical consultant for and then member of the Editorial Board of the New England Journal of Medicine. Dr. Bailar is a member of the International Statistical Institute and was elected to the Institute of Medicine in 1993. He received his MD from Yale University and his PhD in statistics from American University.

Kim Boekelheide is professor in the Department of Pathology and Laboratory Medicine at Brown University. His research interests are in male reproductive biology and toxicology, particularly the potential roles of germ-cell proliferation and apoptosis and local paracrine growth factors in the regulation of spermatogenesis after toxicant-induced injury. Dr. Boekelheide serves on the NRC Subcommittee on Fluoride in Drinking Water and has served on the Committee on Gender Differences in Susceptibility to Environmental Factors: A Priority Assessment. He is a past member of the Board of Scientific Counselors of the National Toxicology Program (NTP) and currently serves on the NTP Center for the Evaluation of Risks to Human Reproduction expert panel that is evaluating di-(2-ethylhexyl)phthalate. Dr. Boekelheide received his MD and PhD (Pathology) from Duke University and is Board Certified in Anatomic and Clinical Pathology.

Robert Brent is Distinguished Professor of Pediatrics, Radiology, and Pathology at the Jefferson Medical College of Thomas Jefferson University and head of the Laboratory of Clinical and Environmental Teratology at the Alfred I. duPont Hospital for Children. Dr. Brent's research focuses on the environmental and genetic causes of congenital malformations, genetic disease, and cancer, with an emphasis on reproduction and the toxicity of drugs, physical agents, and chemicals. Dr. Brent recently completed "The Vulnerability and Resiliency of the Developing Embryo, Infant, Child and Adolescent to the Effects of Environmental Chemicals, Drugs and Physical Agents as Compared to Adults" for the EPA and the American Academy of Pediatrics, which was published in Pediatrics in April 2004. He is a member of the Institute of Medicine. Dr. Brent received his MD with honors; a PhD in embryology, radiation biology, and physics; and an honorary DSc—all from the University of Rochester.

Gail Charnley is principal of HealthRisk Strategies, her consulting practice in Washington, DC. Her interests are toxicology, environmental health risk assessment, and risk-management science and policy. During its tenure, she was executive director of the Presidential/Congressional Commission on Risk Assessment and Risk Management, mandated by Congress to evaluate the role that risk assessment and risk management play in federal regulatory programs. Before her appointment to the commission, she served as director of the toxicology and risk assessment program at the National Academies. She has been the project director for several NRC committees, including the Committee on Risk Assessment Methodology and the Complex Mixtures Committee, and served as the chair of several U.S. Army Science Advisory Board committees that evaluated health risk assessment. Dr. Charnley serves on the NRC Committee on Improving Practices for Regulating and Managing Low-Activity Radioactive Waste. She received her PhD in toxicology from the Massachusetts Institute of Technology.

Vivian G. Cheung is associate professor in the Department of Pediatrics and Genetics at the University of Pennsylvania School of Medicine and a member of the Cell and Molecular Biology and Genomics and Computational Biology Graduate Groups. Her primary research interests include human genome variation, DNA damage repair, and the use of genome-wide approaches to study the genetic basis of human phenotypes and traits. Her research techniques include genomic-mismatch scanning, sequence-mismatch detection, physical mapping, molecular fingerprinting, DNA microarrays, fluorescent image analysis, and developing genome databases. She earned her MD from Tufts University.

Sidney Green, Jr., is graduate professor of pharmacology at Howard University College of Medicine. Dr. Green's research interests include tissue culture, scientific and policy issues associated with alternatives, use of animals in toxicology, and mutagenic assay systems. He has served on the editorial boards of several scientific journals, and he is a fellow of the Academy of Toxicological Sciences. Dr. Green is a member of the NRC Committee on Toxicology and has served on several NRC panels, including the Subcommittee on Acute Exposure Guideline Levels, the Subcommittee on the Toxicity of Diisopropyl Methylphosphonate, and the Subcommittee on Iodotrifluoromethane. He received his PhD in biochemical pharmacology from Howard University.

Karl T. Kelsey is professor of cancer biology and environmental health in the Departments of Genetics and Complex Diseases and Environmental Health at the Harvard School of Public Health. Dr. Kelsey's research interests are in occupational and environmental disease, including susceptibility to disease, with emphasis on gene-environment interactions in the production of chronic disease, and the determinants of somatic gene inactivation in lung and upper airway cancers. He has been at the Harvard School of Public Health since 1987. Dr. Kelsey has served on numerous NRC committees, including the Committee on Copper in Drinking Water, the Committee to Review the Health Consequences of Service During the Persian Gulf War, and the Committee on the Health Effects of Mustard Gas and Lewisite. Dr. Kelsey received his MD from the University of Minnesota and an MOH from Harvard University.

Nancy I. Kerkvliet is a professor in the Department of Environmental and Molecular Toxicology at Oregon State University (OSU). Dr. Kerkvliet also serves as the Associate Director of the Environmental Health Sciences Center at OSU as well as Director of the Flow Cytometry and Cell Sorting Facilities Core. Her research interests include the use of animal models to understand how chemicals alter immune function, particularly the mechanisms of action of polychlorinated dibenzo-p-dioxins and other aryl hydrocarbon receptor (AhR) ligands. Transgenic and gene-deletion approaches are being used, as well as genomics, to address mechanisms of AhR-mediated immunotoxicity. She is also active in public outreach education programs in toxicology and risk communication. Dr. Kerkvliet is a current member of the IOM Committee to Review the Health Effects in Vietnam Veterans of Exposure to Herbicides and is a past member of the NRC's Committee on Toxicology. She has also served as a Councilor for the Society of Toxicology. She earned her PhD degree in interdisciplinary biological sciences and toxicology from OSU.

Abby A. Li recently joined Exponent as a managing scientist/ toxicologist in the Health Risk and Food and Chemical Practices. Her fields of research include toxicology, neurotoxicology, developmental neurotoxicology, psychopharmacology, risk assessment, and pesticide regulation. Previously, Dr. Li was a senior science fellow and a global regulatory science manager at Monsanto, providing expertise in toxicology and risk assessment to address regulatory scientific issues in different world areas. For more than 10 years, she led the neurotoxicology group at Monsanto's Environmental Health Laboratory, where she conducted pharmacokinetic, toxicology, and neurotoxicology studies of industrial chemicals, agricultural products, and pharmaceuticals. Dr. Li served on the U.S. expert teams to the Organisation for Economic Cooperation and Development (OECD) for the development of international test guidelines for adult and developmental neurotoxicology and as chair of neurotoxicology expert groups for industry trade organizations (the American Chemistry Council's long-range research program and the American Industrial Health Council) addressing scientific regulatory issues in neurotoxicology. Dr. Li was a member of the EPA's Science Advisory Board's Environmental Health Committee for 6 years, reviewing the lead rule, 1,3-butadiene risk assessment, trichloroethylene risk assessment, cancer guidelines, the IRIS database, development of acute reference exposure, methods for derivation of inhalation reference concentrations, and indoor air toxics priority-ranking. She is a member of the International Life Science Institute Agricultural Chemical Safety Assessment panel involved in redesign of safety assessment of pesticides. She received her PhD in pharmacology and physiology from the University of Chicago.

George Lucier is a consultant in toxicology. He is a senior adjunct scientist for Environmental Defense, an adviser to the NIEHS and the National Toxicology Program (NTP), and a member of the EPA's Science Advisory Board. He retired from NIEHS in 2000 where he was director of the Environmental Toxicology Program, associate director of the NTP, and head of a research group in molecular epidemiology and dosimetry. In his NTP role, Dr Lucier was responsible for coordinating toxicologic research and testing across federal agencies including EPA, the FDA and the NIOSH. His research focused on the use of basic biology to reduce uncertainty in human risk assessments and to improve the tools used in exposure assessment. Dr Lucier was editor of Environmental Health Perspectives for 28 years and is still a consulting editor. He received his PhD from the University of Maryland School of Agriculture.

Lawrence McCray is a research affiliate with the Massachusetts Institute of Technology (MIT), where he leads a project on the use of knowledge in decision-making and participates in other research on organizational performance and behavior in risk management. Dr. McCray was a staff director and a senior manager at the National Research Council, where he led many studies on U.S. science and technology policy programs, including the seminal study Risk Assessment in the Federal Government: Managing the Process, the so-called Red Book. Dr. McCray also served as head of the EPA Regulatory Reform Unit and a program director on regulatory reform in the Executive Office of the President. He earned a PhD in science and public policy from MIT and an MBA from Union College.

Otto Meyer is head of the Section of Biology, Department of Toxicology and Risk Assessment, Danish Institute for Food and Veterinary Research. The section has overall responsibility for in vivo testing in the department, including repeated dose-toxicity studies, carcinogenicity studies, reproductive-toxicity studies, and neurotoxicity studies. He is the specialized expert to the European Economic Community on classification and labeling of dangerous substances with carcinogenic, mutagenic, or teratogenic properties and national coordinator of the Test Guideline Programme (human health) of the OECD. Concerning the latter commitment, Dr. Meyer is a member of the group preparing an OECD guidance document on reproductive toxicity and assessment. During the last 5 years, he has served as a member of the European Union Scientific Committee on Plant Protection Products (now named the Panel of Plant

Health), Plant Protection and their Residues under the European Food Safety Authority. Dr. Meyer earned a DVM from the Royal Veterinary and Agricultural University in Copenhagen.

D. Reid Patterson retired in 2003 after almost 20 years of responsibility for the toxicity and safety assessment of the diverse portfolio of pharmaceutical, diagnostic, and hospital products for Abbott Laboratories; he is now a private consultant. During his tenure, he led the research efforts in toxicology, pathology, laboratory animal medicine, metabolism, pharmacokinetics, and analytic chemistry in an effort to characterize product hazards. Environmental toxicity was a greater focus during his earlier years in the petrochemical industry (Shell) and the contract laboratory business (Hazleton). Dr. Patterson is a veterinarian with residency training in laboratory animal medicine, and he received his PhD in comparative pathology from the University of Missouri. He is board-certified in laboratory animal medicine, veterinary pathology, and general toxicology, and he is a fellow of the Academy of Toxicological Sciences and the International Academy of Toxicologic Pathology.

William Pennie is director of molecular and investigative toxicology with Pfizer Inc. Dr. Pennie's research interests began with the molecular biology of the estrogen receptor, particularly differential transcriptional regulation by estrogen-receptor subtypes. More recently, his interests have included global receptor biology, improving the predictivity of investigative techniques used at early stages of product development, the technology and application of custom microarray toxicogenomics platforms, and the application of state-of-the-art molecular profiling techniques to research and investigative toxicology. He chairs the International Life Sciences Institute/Health and Environmental Sciences Institute (ILSI/HESI) Committee on the Application of Genomics to Mechanism-Based Risk Assessment. Dr. Pennie received his PhD from the Beatson Institute for Cancer Research at the University of Glasgow, Scotland.

Robert A. Scala is former senior scientific adviser at Exxon Biomedical Sciences Inc. He is also an adjunct professor of toxicology at Rutgers University. He is well known for his work on the toxicity of gasoline components and chemical mixtures. He is a past president of the Society of Toxicology and the American Board of Toxicology. He has published in chronic toxicity testing and evaluation of alternative test protocols and data. Dr. Scala has served on several NRC committees including the Committee on Environmental Justice: Research, Education, and Health Policy Needs, the Committee on Lead Toxicity, and the Committee on Methods for In Vivo Toxicity Testing of Complex Mixtures from the Environment. Dr. Scala earned his PhD in physiology from the University of Rochester School of Medicine and Dentistry.

Gina M. Solomon is a senior scientist at the Natural Resources Defense Council (NRDC) and an assistant clinical professor of medicine at the University of California, San Francisco (UCSF), where she is also the Associate Director of the UCSF Pediatric Environmental Health Specialty Unit. Her work has included research on asthma, pesticides, and environmental and occupational threats to reproductive health and child development. Dr. Solomon serves on the EPA Science Advisory Board Drinking Water Committee and previously served on the Endocrine Disruptor Screening and Testing Advisory Committee. Dr. Solomon received her MD from Yale University and underwent her postgraduate training in medicine and public health at Harvard University.

Martin Stephens is vice president of the Animal Research Issues Section of the Humane Society of the United States. Dr. Stephens serves as coordinator of the International Council for Animal Protection at the OECD. He also serves on the Scientific Advisory Committee on Alternative Toxicological Methods for the National Toxicology Program Interagency Center for the Evaluation of Alternative Toxicological Methods and on the Scientific Advisory Panel of the Institute for In Vitro Sciences. Dr. Stephens has extensive experience in animal protection and in vitro testing sciences. He earned a PhD in biology from the University of Chicago.

James Yager is professor of toxicology in the Department of Environmental Health Sciences, director of the NIEHS Training Program in Environmental Health Sciences, and senior associate dean for academic affairs at the Johns Hopkins University Bloomberg School of Public Health. Dr. Yager is a member and a past president of the carcinogenesis specialty section of the Society of Toxicology. His research focuses on the role of catechol metabolites of endogenous, synthetic, and environmental estrogens and polymorphisms in genes involved in estrogen metabolism as risk factors in the development of cancer of the breast and liver. Dr. Yager earned his PhD from the University of Connecticut.

Lauren Zeise is chief of the Reproductive and Cancer Hazard Assessment Branch of the California Environmental Protection Agency. She received her PhD from Harvard University. Dr. Zeise's research focuses on modeling human interindividual variability and risk. She has served on advisory boards of the U.S. EPA, the World Health Organization, the Office of Technology and Assessment, and the NIEHS. She has also served on several NRC committees, including the Committee on Risk Characterization, the Committee on Comparative Toxicology of Naturally Occurring Carcinogens, the Committee on Copper in Drinking Water, and the Committee to Review EPA's Research Grants Program. Dr. Zeise is a member of the Board on Environmental Studies and Toxicology and of the Institute of Medicine's Health Promotion and Disease Prevention Board.

Appendix B

Testing Protocols

OPPTS HARMONIZED TESTING GUIDELINES[1]

Series 870 Health Effects

870.1000 Acute toxicity testing-background
870.1100 Acute oral toxicity
870.1200 Acute dermal toxicity
870.1300 Acute inhalation toxicity
870.2400 Acute eye irritation
870.2500 Acute dermal irritation
870.2600 Skin sensitization
870.3050 Repeated dose 28-day oral toxicity study in rodents
870.3100 90-Day oral toxicity in rodents
870.3150 90-Day oral toxicity in nonrodent
870.3200 21/28-Day dermal toxicity
870.3250 90-Day dermal toxicity
870.3465 90-Day inhalation toxicity
870.3550 Reproduction/developmental toxicity screening test
870.3650 Combined repeated dose toxicity study with the reproduction/
      developmental toxicity screening test
870.3700 Prenatal developmental toxicity study

---

[1]The EPA OPPTS Harmonized Guidelines can be found online at
http://www.epa.gov/opptsfrs/home/guidelin.htm.

870.3800 Reproduction and fertility effects
870.4100 Chronic toxicity
870.4200 Carcinogenicity
870.4300 Combined chronic toxicity/carcinogenicity
870.5100 Bacterial reverse mutation test
870.5140 Gene mutation in Aspergillus nidulans
870.5195 Mouse biochemical specific locus test
870.5200 Mouse visible specific locus test
870.5250 Gene mutation in Neurospora crassa
870.5275 Sex-linked recessive lethal test in Drosophila melanogaster
870.5300 In vitro mammalian cell gene mutation test
870.5375 In vitro mammalian chromosome aberration test
870.5380 Mammalian spermatogonial chromosomal aberration test
870.5385 Mammalian bone marrow chromosomal aberration test
870.5395 Mammalian erythrocyte micronucleus test
870.5450 Rodent dominant lethal assay
870.5460 Rodent heritable translocation assays
870.5500 Bacterial DNA damage or repair tests
870.5550 Unscheduled DNA synthesis in mammalian cells in culture
870.5575 Mitotic gene conversion in Saccharomyces cerevisiae
870.5900 In vitro sister chromatid exchange assay
870.5915 In vivo sister chromatid exchange assay
870.6100 Acute and 28-day delayed neurotoxicity of organophosphorus
        substances
870.6200 Neurotoxicity screening battery
870.6300 Developmental neurotoxicity study
870.6500 Schedule-controlled operant behavior
870.6850 Peripheral nerve function
870.6855 Neurophysiology Sensory evoked potentials
870.7200 Companion animal safety
870.7485 Metabolism and pharmacokinetics
870.7600 Dermal penetration
870.7800 Immunotoxicity
870.8355 Combined Chronic Toxicity/Carcinogenicity Testing of
        Respirable Fibrous Particles

OECD TESTING GUIDELINES

Series 4: Health Effects

401 Acute Oral Toxicity (Deleted Guideline, date of deletion: 20th December 2002)

402 Acute Dermal Toxicity (Updated Guideline, adopted 24th February 1987)

403 Acute Inhalation Toxicity (Original Guideline, adopted 12th May 1981)

404 Acute Dermal Irritation/Corrosion (Updated Guideline, adopted 24th April 2002)

405 Acute Eye Irritation/Corrosion (Updated Guideline, adopted 24th April 2002)

406 Skin Sensitisation (Updated Guideline, adopted 17th July 1992)

407 Repeated Dose 28-day Oral Toxicity Study in Rodents (Updated Guideline, adopted 27th July 1995)

408 Repeated Dose 90-Day Oral Toxicity Study in Rodents (Updated Guideline, adopted 21st September 1998)

409 Repeated Dose 90-Day Oral Toxicity Study in Non-Rodents (Updated Guideline, adopted 21st September 1998)

410 Repeated Dose Dermal Toxicity: 21/28-day Study (Original Guideline, adopted 12th May 1981)

411 Subchronic Dermal Toxicity: 90-day Study (Original Guideline, adopted 12th May 1981)

412 Repeated Dose Inhalation Toxicity: 28-day or 14-day Study (Original Guideline, adopted 12th May 1981)

413 Subchronic Inhalation Toxicity: 90-day Study (Original Guideline, adopted 12th May 1981)

414 Prenatal Developmental Toxicity Study (Updated Guideline, adopted 22nd January 2001)

415 One-Generation Reproduction Toxicity Study (Original Guideline, adopted 26th May 1983)

416 Two-Generation Reproduction Toxicity Study (Updated Guideline, adopted 22nd January 2001)

417 Toxicokinetics (Updated Guideline, adopted 4th April 1984)

418 Delayed Neurotoxicity of Organophosphorus Substances Following Acute Exposure (Updated Guideline, adopted 27th July 1995)

419 Delayed Neurotoxicity of Organophosphorus Substances: 28-day Repeated Dose Study (Updated Guideline, adopted 27th July 1995)

420 Acute Oral Toxicity - Fixed Dose Method (Updated Guideline, adopted 20th December 2001)

421 Reproduction/Developmental Toxicity Screening Test (Original Guideline, adopted 27th July 1995)

422 Combined Repeated Dose Toxicity Study with the Reproduction/Developmental Toxicity Screening Test (Original Guideline, adopted 22nd March 1996)

423 Acute Oral toxicity - Acute Toxic Class Method (Updated Guideline, adopted 20th December 2001)

424 Neurotoxicity Study in Rodents (Original Guideline, adopted 21st July 1997)

425 Acute Oral Toxicity: Up-and-Down Procedure (Updated Guideline, adopted 20th December 2001)

427 Skin Absorption: In Vivo Method (Original Guideline, adopted 13th April 2004)

428 Skin Absorption: In Vitro Method (Original Guideline, adopted 13th April 2004)

429 Skin Sensitisation: Local Lymph Node Assay (Updated Guideline, adopted 24th April 2002)

430 In Vitro Skin Corrosion: Transcutaneous Electrical Resistance Test (TER) (Original Guideline, adopted 13th April 2004)

431 In Vitro Skin Corrosion: Human Skin Model Test (Original Guideline, adopted 13th April 2004)

432 In Vitro 3T3 NRU Phototoxicity Test (Original Guideline, adopted 13th April 2004)

451 Carcinogenicity Studies (Original Guideline, adopted 12th May 1981)

452 Chronic Toxicity Studies (Original Guideline, adopted 12th May 1981)

453 Combined Chronic Toxicity/Carcinogenicity Studies (Original Guideline, adopted 12th May 1981)

471 Bacterial Reverse Mutation Test (Updated Guideline, adopted 21st July 1997)

473 In vitro Mammalian Chromosomal Aberration Test (Updated Guideline, adopted 21st July 1997)

474 Mammalian Erythrocyte Micronucleus Test (Updated Guideline, adopted 21st July 1997)

475 Mammalian Bone Marrow Chromosomal Aberration Test (Updated Guideline, adopted 21st July 1997)

476 In vitro Mammalian Cell Gene Mutation Test (Updated Guideline, adopted 21st July 1997)

477 Genetic Toxicology: Sex-Linked Recessive Lethal Test in Drosophila melanogaster (Updated Guideline, adopted 4th April 1984)

478 Genetic Toxicology: Rodent Dominant Lethal Test (Updated Guideline, adopted 4th April 1984)

479 Genetic Toxicology: In vitro Sister Chromatid Exchange Assay in Mammalian Cells (Original Guideline, adopted 23rd October 1986)

480 Genetic Toxicology: Saccharomyces cerevisiae, Gene Mutation Assay (Original Guideline, adopted 23rd October 1986)

481 Genetic Toxicology: Saacharomyces cerevisiae, Miotic Recombination Assay (Original Guideline, adopted 23rd October 1986)

482 Genetic Toxicology: DNA Damage and Repair, Unscheduled DNA Synthesis in Mammalian Cells in vitro (Original Guideline, adopted 23rd October 1986)

483 Mammalian Spermatogonial Chromosome Aberration Test (Original Guideline, adopted 21st July 1997)

484 Genetic Toxicology: Mouse Spot Test (Original Guideline, adopted 23rd October 1986)

485 Genetic Toxicology: Mouse Heritable Translocation Assay (Original Guideline, adopted 23rd October 1986)

486 Unscheduled DNA Synthesis (UDS) Test with Mammalian Liver Cells in vivo (Original Guideline, adopted 21st July 1997)